# Walther Cloos

# Gemstones and Earth's History

WALTHER CLOOS

# GEMSTONES AND EARTH´S HISTORY

Formative Forces Publishing

2021

Originally Published by
Verlag Freies Geistesleben, Stuttgart 1956
*Kleine Edelsteinkunde: im Hinblick auf die Geschichte der Erde*

Translated by Jannebeth Röell and James Lee

ISBN Printed Book: 978-1-7365513-2-5

ISBN E-Book: 978-1-7365513-3-2

Library of Congress Control Number: 2021914189

Cover Image: Quartz, Tibet
(Cover photo by JJ Harrison, Courtesy of Wikimedia Commons) [i]

To the memory

of our friend and geographer

Otto Emanuel Eckstein

Born Stuttgart, March 22, 1894

Crashed on the Piz Rosegg in the Bernina

September 5, 1944

## Contents

## Translators' Foreword

With "Gemstones and Earth's History," Walther Cloos tells the story of gemstones as they develop through earth's history. His love for the earth and its gemstones, minerals, and metals shine through it while offering readers the opportunity to understand how gemstones and their parent minerals came into being. Throughout the book, the author explores the relationship of minerals, metals, and gemstones to the human organism. His insights into the ethereal aspects of nature immediately deepen the reader's insight into the world's origins. He aligns interesting events and the formation of substances from the earth's oldest past with current observations.

Each of the original mineral and gemstone names used in this text has been checked against modern mineral atlases and other sources to ensure they can easily be looked up online or found in a library. Those terms found to be archaic, outdated, ambiguous, or no longer in use were changed to reflect modern usage.

Similarly, many place names and country boundaries have changed since the book was originally written and published in 1956. Current place names are used where this is the case.

Some of the original figures in the book were replaced with higher-quality examples.

These are the only changes made to the author's original work. The translators have worked diligently to keep the author's "voice" present throughout the book.

Additional information regarding the minerals and gemstones named in this book can be found at www.mindat.org and www.mineralatlas.eu, among other mineral databases and information sources on the Internet.

We hope you will find this work a worthwhile read.

## Preface

This book originates from the desire of a group of people to know something about gemstones not found in scientific literature. They sought new ways to observe natural phenomena and natural forms and found it in Rudolf Steiner's life work, anthroposophy.

In response to this desire, the author has here attempted to expose the observable facts to the light of Rudolf Steiner's spiritual-scientific research concerning the true essence of the mineral realm. From these points of view, something is expressed in the gemstones about the whole of nature and the human being, whose essential foundation has not yet been realized. The gemstones are creation's pointer to a realm of world existence that we will approach in the distant future. That is why today, every attempt to take the spiritual image of human beings and nature seriously is condemned to incompleteness.

The author nevertheless found the courage to attempt it with the excuse that he wanted to point to the beauty of the gemstones. This beauty is a gateway that can lead into an unknown world: "only through the dawn of beauty can you enter the realm of knowledge!"[1]

Stuttgart, Johnstide, 1956

Walther Cloos

---

[1] Rudolf Steiner, *The Nature and Origin of the Arts: The Being of the Arts*, Berlin, October 28, 1909, GA 271.

Figure 1: Rock crystal rhyton from Heraklion, Greece. Minoan, 1500-1450 BC. The vase has an elongated body with a pointed bottom made of an unusually large piece of rock crystal. The neck and splaying rim, also of rock crystal, are joined to the body by a crystal ring decorated with vertical rings of gilt faïence. The tall handle is formed by fourteen beads of gilt crystal threaded onto a bronze wire. (Photo Courtesy of Wikimedia Commons)[ii]

## 1. Human Beings and Gemstones

Anyone who has walked in the mountains and had the good fortune to find a clear rock crystal in a crevice knows the astonishment at the sight of this somehow otherworldly formation. Something about the clarity in the air, something about the transparent purity of bubbling springs, seems to be connected with the rigor that rules the course of the stars. The dark density of matter seen in every common stone is as if shooed away by a mysterious power, which has now conjured up the crystal from the blue-black schist or dense, impenetrable granite.

When the unbiased person returns home from a long walk and asks about the essence of crystals and gemstones in the many books written on the subject, not much from this enchantment remains. They find fantastic formulas about chemical composition and strict mathematics regarding the crystal forms of the stones. They find profound theories about presumed chemical and physical processes that led to these pure and austere formations, which emerged in the distant past from the glowing primordial magma.

Yet nothing is said about the undeniable miracle embedded in the dense, opaque rock, where clear, water-bright, and colorful crystals are found as if floating. Also not mentioned are the great rock crystals of two meters in length and fifty centimeters in diameter that were not formed in the most ancient times but in our earth's recent mountain formations. Many such wonders are encountered by those who study the origin and localities of these gemstones. Indeed, these wonders are described in the books with all the care that is proper for scientific research. In the course of our considerations about individual gemstones, we will get to know many such characteristics. However, these will mean nothing to us if we do not learn to awaken and train our soul force of wonder. Wonder, devotion, and humility are the foundations of new powers of cognition to which Rudolf Steiner has repeatedly referred. They are valid for all research and have their special validity for the chaste, pious world of mineral formations.

The scientific thinking of the last centuries has brought us a comprehensive knowledge of nature, which was necessary to expand our consciousness and sharpen our senses. But with this expansion of our consciousness and knowledge of outer nature, we have almost completely lost the inner essence of nature, which today, as in the past, can be the true source of our wonder and awe.

In humanity's earliest times, in the early cultures of Asia, Africa, and the Americas, when priestly kings still led their people and religious groups, gemstones were in the hands of priests and kings alone. Their knowledge of the relation between natural phenomena and the world's creative forces allowed them to find the places, even in the remotest areas, in bedrock or the debris of river valleys and mountain slopes, where those wonderful gems rested. For them, they represented the highest expression of divine activity in the dead world of stone. In meditation, the color and form of the crystal led their gaze out into the vastness of the zodiacal sky, where they saw those world-creating beings whose work for the evolution of human beings and the earth had also created the gemstones. That is why since the oldest times, gemstones were related to the zodiac. In his writings, Agrippa von Nettesheim communicated the tradition of the Jewish Kabbalah and the Greek Mysteries regarding gemstones and their relation to the zodiac signs. On these traditions rest the more or less questionable assignments of stones to a month, with which a businesslike industry accommodates the mystical needs of modern human beings without knowing that such assignments carry real secrets. This desecration marks such "assignments" as false and insubstantial.

Such desecration was completely impossible in those past times. The gemstone was regarded as the highest expression of divine activity in the dead stone world; it was considered sacred and could not be in anyone's personal possession. The priest and king who wore the gemstone as a ring, or wrapped in precious fabric, or as an emblem of the crown, expressed to the people their intimate contact with the divine powers of the world and that they, when clothed with these insignia, were only a servant of these powers. In later times, when we see these gemstones as an expression of the wealth and power possessed by kings and priests, we are already dealing with signs of decadence that indicate that the original sense for the divine beings associated with these jewels has been lost.

Despite this decadence, a knowledge of the oldest secrets of gemstones has been preserved by certain groups of people until relatively recent times. This knowledge was expressed where gemstones were used for the adornment of churches and smaller church rooms. This precious material can be seen most clearly in the Chapel of St. Wenceslas in St. Vitus Cathedral in Prague (built by Peter Parler) and in the Chapel of the Holy Cross at the old Karlstein Castle near Prague. In these two buildings, the chapel space is lined all around with irregular, polished slabs of rock crystal, rose quartz, amethyst, chrysoprase, and other gemstones from the mountains of Germany on its eastern border. The slabs have a size up to 25 by 25 centimeters. The joints between the slabs are lined with real gold. In the chapel at the Karlstein Castle, there is said to have been a whole window made of cut and set gemstone tiles.

When a person enters such a chapel, they do not have the impression of overwhelming splendor. The delicate transparency of the gemstone walls only enhances the solemnity and dignity of the space, which nevertheless seems to be subdued by an unreal light from above through the colored glass windows. The actual beauty of these gemstones, showing themselves by the light shining through them, does not come into its own at all in this type of use.

A similar "indifference" toward the splendor of gemstones can also be seen where they are used as "jewelry" and "ornamentation" on ancient religious altars, relics, and sarcophagi. The large and valuable stones used for this purpose are not, as is generally the case today, artfully cut with many light-refracting facets but are merely rounded and polished with the greatest possible respect for their natural form. In this way, only the color and pure translucency of the stone is shown to advantage, and the eye is not "dazzled." Thus, the large, gilded sarcophagus of the

Chapel of St. Wenceslas in St. Vitus Cathedral in Prague is set with rubies, sapphires, and emeralds worked in such a simple manner. If you look around in the older church treasuries, you will find that only recently, at the end of the 15th century, at the beginning of the modern age, a new era began for the processing of gemstones.

Figure 2a: Irregular, polished gemstone slabs in the walls of the private chapel of Charles IV, Karlsteijn Castle.

Figure 2b: Closeup of the irregular, polished gemstone slabs in the walls of the private chapel of Charles IV, Karlsteijn Castle.

This simple way of processing stones is still common in India for gemstones found there. It was never based on a lack of technical knowledge because today, the great art of polishing the stones is still practiced without mechanical aids by the Indian people. The once exclusive use of gemstones for cultic purposes, completely renouncing the "dazzling splendor" of faceted stones, is connected with a secret, the revelation of which we owe to Rudolf Steiner. To better understand this riddle, it will be good to realize what it actually means to have transparent material, meaning matter that is permeable to light. In the formation of clouds, we notice that air and water are transparent, water much less so than air, which seems self-evident. It seems natural to relate these facts to density, that is, the matter's specific gravity. Air is lighter than water; water is lighter than rock. Roughly speaking, the transparency seems to be related to the density of the materials. Therefore, the gemstones should not be much heavier than water. Nevertheless, we note that gemstones have on average two to five times the specific gravity of water, and in some cases, they reach the specific gravity of metals. Thus, we cannot relate the transparency of gemstones to density. However, observation shows us that the transparency can be of the purest "water." (For diamonds and other colorless stones, the term "water" is used in the technical qualitative grading of the stone). If one examines gemstones for possible water content, it shows that just the ones of purest "water" contain no trace of it.

Many gemstones, especially if they are cut with facets, show another phenomenon called the "fire" of the stone. The fire of a stone is related to the way it refracts and reflects incident light. The angle of refraction depends on the stone's specific gravity, in other words, on the stone's density. Gemstones with higher specific gravity, such as diamond, hyacinth (zircon), ruby, sapphire, and garnet, show stronger refraction or even double refraction than, for example, the much lighter rock crystal. Here we also discover that the fire of a stone increases with its "weight." This contradicts all experiences that we otherwise have with the matter because the heaviest substances of the earth, the metals, are completely impervious to light. We also see that our usual notions of density and gravity, water and fire, do not apply to gemstones. They are simply a form of matter which represents a special case.

We already mentioned at the beginning that this "special case" consists of the fact that a gemstone can appear in the middle of an opaque rock. If we look for a comparison, a phenomenon that shows similarities in a completely different area, then we would have to think of the emergence of the eye formation, where the translucent substance of the eyes lies in the opaque mass of the body, embedded in the bone cavities of the skull. Such a comparison may seem daring because it seems the formation of gemstones has little to do with human eyes, although it may occur to a poet to compare the eyes of their beloved to gemstones.

If the poets and artists had had a say in science, this attempt could have been the starting point for an insight, which we owe to the research of Rudolf Steiner. We would not have dared to present this comparison if we had not had his indications of an inner connection between the development of the sense organs and the formation of the gemstones of the earth. We will come again and again to this connection when considering the individual gemstones.

The work of Lili Kolisko to show the influence of silica on plant growth has demonstrated that transparent matter, such as quartz sand (rock crystal), has relationships to life processes. Lili Kolisko proved that plants also form leaf green in the dark when there is a high quartz sand content in the soil, which otherwise only develops when the plant is exposed to direct light. The silica imparted the effect of light it had stored within itself to the plants growing in the dark. Other experiments have shown that strongly illuminated quartz sand allows green seedlings to grow in the dark, while freshly removed sand unexposed to light produced pale seedlings. Rudolf Steiner's indication in his agricultural lectures that silica is something like a light-warmth sense organ for the whole earth becomes understandable from these experiments. But it is also evident from such phenomena that transparent matter is endowed with properties that cannot be determined by chemical-physical methods. They become apparent only in connection with the life processes.

From what has been said, it may become clear, at least at the beginning, that the transparent matter, which the gemstone expert compares with water and fire, is really a state of a substance that is lifted out of the general rock-being of the earth, like the shining, light-transmitting eye of a human being out of its own body.

The gemstones are not just accidental formations whose origin can be "interpreted" from chemical-physical considerations. They are basically miraculous creations like the human eyes or other human senses with their "permeability" for sound, warmth, form, or spiritual qualities. If, however, other forces enter through our senses than what we can physically perceive as light, sound, warmth, form, and so on, then the idea is not at all so far fetched that the gemstones likewise still exhibit "permeabilities," which at first elude our cognitive capacity.

Here we may come back to the ritual uses of gemstones in past times mentioned above, and perhaps we can now begin to understand that Rudolf Steiner, in larger contexts, has characterized the gemstones as sense organs of high spiritual beings, through which these beings, which have no physical body, are given the possibility to look into earthly physical events. The same world-creating beings, who built up the human organism's miraculous structure through long periods of evolution, also created their sense organs as counter-images in the earth's minerals: the gemstones. This occult fact was known to the priest-kings of ancient times and many other great initiates. It must still have been known by the church builders of the Gothic period. They dressed the rooms of the chapels, altar-utensils, relics, and sarcophagi with gemstones in a simple form so that during the celebration of the ritual, the spiritual world could look in and participate in the activities of human beings.

## 2. The Rock Crystal

In a collection of nice stories known as *Colorful Stones*, Adalbert Stifter gives us a tender Christmas tale entitled "Rock Crystal." In this story, two children who encounter distress in the wintery mountains are saved from death by a miraculous, protective, and enveloping fate. Stifter shows us that if an eye is not only searching but also lovingly and artistically observing the world's phenomena, it can delve deeper into the secrets connected with the world of rocks. It is the fundamental mood of the rock crystal, if one may use this expression here, that envelops and conserves and brings precise enclosed forms of clarity and alertness with it.

The same motif meets us in the fairy tale of Snow White and her glass coffin, which is nothing other than the shape of a rock crystal. Snow White is the soul of the human being enclosed in the body's coffin and kissed awake by the human being's spirit being, the I, the prince.

This artistic and fairy-tale-like interpretation of the rock crystal concept is not based on abstract symbolism but on the scientific fact that the substance of rock crystal, this siliceous substance, plays an important role in the skin surrounding the human body. This silica substance is not simply embedded in the human skin, but it unfolds an *activity* there. While this substance is constantly being formed from within and then shed outwards, in human consciousness, the contour of our own shape can be felt, and simultaneously the connection of the body's boundary with the environment can be perceived. This perception of the surrounding world is now differentiated into the various capacities of the senses of seeing, hearing, touching, smelling, and tasting. All these senses are based on the *activity* of silica at the outer perimeter of the human organization. But just as the general perception of the body's environment through the skin is subdivided into the various senses mentioned above, so also silica is to be found in the world of stones, on the one hand as simple quartz or rock crystal, and on the other hand, subdivided into the various gemstones. Thus there is an evolutionary relationship (according to Rudolf Steiner's lectures of October 1906 and Agrippa von Nettesheim) between:

- the sense of seeing and the formation of the chrysolite,
- the sense of hearing and the formation of the onyx,
- the sense of touch and the formation of the carnelian,
- the sense of smell and the formation of the jasper,
- the sense of taste and the formation of the topaz.

All these gemstones contain silica; however, it is combined with other substances, has a different structure (inner composition), or is formed in a very different way than the simple rock crystal.

Later, when we look at the individual gemstones, we will always return to the original and basic substance of silica. Rudolf Steiner called silica the "universal sense organ in the earth" in his lectures on agriculture.[2] Through it, the earth and the plant world belonging to it receive the radiating power of light and warmth from the cosmic environment. On these paths of light and warmth, the plant forms have come into the earth in the past and have brought the silica substances with them.[3] Today, they are deposited in the siliceous primary rocks of the earth, in the granites, gneisses, schists, and porphyries. Many of these rocks show through their mineral structures peculiar wood- and plant-like forms, such as asbestos, wood-like chrysotile, sepal-like muscovite, and formations that are the last traces of the actual plant origin of these siliceous rocks. The selfless, light- and warmth-mediating essence of silica is contrasted by the sucking, eager calcareous lime, which came to the earth through animal shapes in the form of bones and shells.

---

[2] Rudolf Steiner, *The Agricultural Course*, Koberwitz, Silesia, June 7th to 16th, 1924, Biodynamic Association, 1958.

[3] Rudolf Steiner, *Mystery Centers*, Dornach, Switzerland 23rd November to 23rd December 1923.

If one now examines the earth's richly varied rocks, as far as they are known to us through outcroppings, drillings, and mines, it is possible to see that more than seventy-five percent of all the rock masses consist of silica. But only a very small amount of this widespread silica is to be found as actual rock crystal. Only where the movement of the still soft plastic rock masses of the past allowed fissures, crags, passages, or bubble-like cavities to develop was the gel-like soft and water-containing silica able to collect and later crystallize in these spaces.

Not only did the silica crystallize in such hollow spaces but also in lime, gypsum, and marble. This is only understandable if one knows that lime, marble, and gypsum, *before* their consolidation, were also in a jelly-like, colloidal state, which concurrently contained the dissolved silica. Consequently, the rock crystal could crystallize while suspended in the soft mass and form particularly beautiful doubly terminated crystals. These so-called doubly terminated crystals show the rock crystal in its most perfect form, with a six-sided column bounded by a six-sided pyramid at both ends. Such doubly terminated rock crystals are found embedded in hollow spaces in the marble of Carrara, Italy, from which Michelangelo created his famous marble sculptures. These so-called tears of Carrara belong to the finest and purest rock crystals known, except that their size is usually very limited.

Other sites for doubly terminated crystals are Kalabagh in India, where the crystals are embedded in the gypsum of salt marl, and Lake George in New York State, where the crystals occur in calcareous sandstone. There may be many more such places on earth. In the mid-1930s, the author had the opportunity to hold thousands of small doubly terminated crystals from the cavities of the shell-rich limestone of the upper Muschelkalk near Gross-Sachsenheim in Württemberg. These crystals were water clear and fully formed up to a size of 1 to 2 millimeters. By dissolving the surrounding lime with acid, the crystals were easily extracted.

The formation of such doubly terminated crystals within a materially different environment shows that the chemical forces are strongly receding in the early stages of the nascent rock world. If a rock crystal – meaning silicic *acid* – can be formed within a gel-like mass of calcium carbonate – a *basic* substance – then the crystal-forming, isolating forces must be dominant. Otherwise, the silicic acid would combine *chemically* with the calcium carbonate, forming calcium metasilicate (wollastonite). This touches upon the archetypical phenomenon of crystal formation and gemstone formation in general. For all crystal and gem formation isolates itself, individualizes itself from the rest of the often very complicated mass of the rock.

One gets the impression that a crystal-forming force sucks a pure substance from the rock mass, leaving behind all that is impure. Even from dark, iron-rich, or otherwise colored rocks, water-clear, colorless rock crystals can crystallize. In a few places, small rock crystals are occasionally found that enclose other materials, have taken them into their form, and are thus opaquely colored. Thus one finds iron quartz (eisenkiesel) that contains finely distributed iron oxide in the Sundwig-Iserlohn area of Germany; from Madagascar come crystals that are rust-red and opaque from iron oxide; and, on Sicily, one finds tiny rock crystals with enclosed sulfur. In the silica permeated petrified wood of older strata, small crystals are sometimes found that are completely black from carbon.

The crystal-forming forces are such that they penetrate the earth from all sides of the universe. This circumstance can be seen clearly in doubly terminated crystals, which are formed on all sides. On the other hand, if a crystal develops with one end, it is actually rooted in the earth because its forming forces could only work from one side. According to Rudolf Steiner, these crystal-forming forces should be looked for in the lines and sides that border the crystals and then extend them into the universe. One comes then to certain star constellations, out of which the crystal-forming forces are or were active.

Whatever one might call it, this space or star crystallography is a matter of higher mathematics that we do not want to explore here. But even without mathematics, we can find these omnidirectional crystal-forming forces, especially the "hexagonal force" of the rock crystal formation, as Rudolf Steiner once called it. We must turn our attention to other areas and away from the rock world to accomplish this. When we turn our attention toward processes occurring in the atmosphere, we can see that snow crystals, for example, are also formed according to this principle of the hexagon. And those fine ice needles, which float at great heights on clear winter days, are tiny six-sided columns oriented vertically in the ice-cold air. The light of the sun refracts and creates by reflection those enchantingly beautiful sun halos with secondary suns standing exactly on the cross.

If we descend from the high atmosphere to the plant world, this hexagonal silica power can clearly be found in microscopic plant cells. Even if these hexagons are not as regular as ice crystals, a phenomenon is repeated here on the level of life, which we also find in the rock world. Mica flakes in gneiss and granite are also hexagonal like the plant cells and lie in layers, one on top of the other similar to the plant. Rudolf Steiner compared the inner structure of the plant with the structure of granular rocks. He pointed out that this cellular structure indicates that the plant substance strives toward the mineral. With the formation of wood, this process is then led to an end within the plant realm.

But this "hexagonal force" also meets us in the flowers of many plants. Thus the flowers of the lilies are arranged in a hexagon. From here, it is not far to the realm of the bees, which know how to handle the "hexagonal force" like no other living being. The honeycomb construction for brood and honey is carried out in the best sense of the word from these silica forces, penetrating the world from all sides. Here the "plant cell" of unfathomable wisdom is conjured up loud and clear before the admiring eye of the human being.

This cell and honeycomb structure is also a phenomenon that we meet again in the rock world. But it is concealed there. Many agates (consisting of hydrated silica) reveal a clear cell and honeycomb structure when cut, polished, and placed in the right light. We will come back to this observation when we look at valuable gemstones related to the agate.

The remarkable thing is that this hexagonal formation of the rock crystal does not come about as with hexagonal crystals of other substances, but through many quite differently shaped individual crystals that coalesce to form a six-sided column. Here it becomes clear that the formative forces of the rock crystal simply overtake the original tendencies of the silica substance and produce a so-called *habit* that is not related to the original crystallization process, the original crystallization tendency. Only rarely does one discover from the Carrara marble such "original" silica crystal forms. They are often only tiny crystals that remind one of everything but the form of the rock crystal. They are embedded in the dense marble and can be extracted only if the marble is immersed in hydrochloric acid.

The habit of the rock crystals mentioned above, which comes about through accumulation and twinning, is now "individualized" in rock crystals over the entire earth. A person with a good memory for forms, who has discovered many rock crystals from numerous places, can easily look at a rock crystal and tell where it originates. This different appearance of the rock crystals has a certain lawfulness characteristic of specific places of origin. Therefore, this lawfulness has been named after places of discovery without explicitly stating that the form in question occurs exclusively at that place.

Variety within this lawfulness occurs because the "original single crystals," which have a rhombohedral or trapezohedral character, can coalesce "right-handed" or "left-handed." In this way, "right-handed crystals" and "left-handed crystals" are formed. Furthermore, *one* right-handed crystal or *one* left-handed crystal can never form a rock crystal on its own. Only when two right- or two left-handed quartz crystals, or when one right-handed and one left-handed quartz crystal complement each other, interpenetrate and coalesce as twins, does a complete rock crystal form. The first case, where two right-handed quartz crystals or two left-handed quartz crystals intertwine, corresponds to the so-called Dauphiné law because it is mainly seen in crystals of the Alps and the Dauphiné. The second case, where a right-handed crystal and a left-handed crystal are twinned, corresponds to the so-called Brazilian law, which takes its name because this law is observed mainly in the crystals of Brazil. However, this Brazilian law is also observed in alpine quartz crystals.

A third case results from a combination of the Dauphiné law and the Brazil law and gives rise to so-called quadruplets. The fourth, the so-called Japan law, which arises because of some excellent occurrences in Japan, comes about when the twins do not interpenetrate in the vertical axis, but the axes are quite angularly inclined to each other. Such Japan law crystals are remarkably flat. In addition, six rarer laws are characteristic of certain little-known localities. They are mentioned because the multitude of possibilities shows the complexity and variability of rock crystal formation. Some alpine quartz crystals (mainly so-called smoky quartz crystals, which we will discuss later) exhibit a "twisted" form. This is caused by the fact that the vertical axes of twins are slightly inclined to each other. Two right-handed crystals can coalesce to a right-handed crystal and two left-handed crystals accordingly. Finally, many crystals show an uneven flat extension and strong distortion.

Without natural visuals or good illustrations, such an enumeration of laws naturally remains very abstract. But perhaps it gives a small idea of the variety of possibilities and shows how the formative forces of crystals vary over the earth so that one can even define their diversity mathematically. But what can be abstractly defined mathematically is the expression of that diversity. This expression of variety results from the crystal-forming forces from the stars seizing earthly substances in diverse locations. These locations of the living earth are very different because of those forces that are expressed as invisible, etheric ones, like those of the plant world. This plant world also originates from the cooperation between the cosmos and the earth. It is so diverse because of the changing heavens above the earth and other forces emanating from inside of the earth. Today, that which still makes the plant world so different over the entire earth made this *apparently* dead world diverse in the past, when the rock world emerged from the living earth.

Figure 3: Doubly terminated quartz crystal from the East Coleman Mine, Jessieville, Arkansas. (Photo Rob Lavinsky, Courtesy of Wikimedia Commons)[iii]

The main country that supplies many rock crystals today is Brazil. In Goyaz, the crystals are found in a hard sandstone in veins and pockets. These crystals are characterized by a slender, narrow habit and are somewhat thicker where they expand at the base than at the top. Their thickness can reach 50 centimeters, their length up to 70-80 centimeters. Also, larger doubly terminated crystals and the so-called phantom quartz crystals are known from Brazilian localities. The phantom quartz crystals show in a specific way *how* a rock-crystal is formed. One sees *in* the transparent crystal parallel to the outer edges and faces 2, 3, and up to 40 smaller crystal faces, which seem to float in a ghostly manner and only show dull, weakly contoured faces in the interior. The opaque surrounding areas of these nested crystals are formed by tiny gas pockets or fluid inclusions. The whole thing looks as if several crystals were nested inside each other. This clearly shows that the rock crystal has really "accumulated," it has gradually and rhythmically accrued new material. And between the individual "periods" of accumulation, a layer of gas or fluid has settled on the surface of the finished crystal, which now shows us the contours of the enclosed crystals.

Giant mountain crystals were found as early as the 18th century in mighty "crystal caverns" on the Grimsel in the Bernese Oberland and in the Viesch Valley in the Upper Valais. Another phenomenon that throws light on the secret of crystal formation was studied with these crystals: many of the crystals found in these cavities were more than one meter in size with corresponding diameters. For example, one crystal had a respectable size of 2.5 meters and weighed 363 kilograms. However, the peculiarity of these crystals was tiny inclusions in *layers*, similar to the phantom quartz crystals, with the difference that these layers were not parallel to the outer crystal faces but formed horizontal zones. These horizontally layered zones ran at the same height through all the crystals, which stood on the crystal cavern's floor. If, for example, such a layer was 50 centimeters from the floor, this layer ran through all other crystals at the same height of 50 centimeters. The next layer could be 56 centimeters above the floor, another 63 centimeters, and the crystals next to it would always have their layer at the same height. This phenomenon is described in the mineralogical literature of the nineteenth century. From this phenomenon, it appears that the whole crystal cavern must have been originally filled with a gelatinous mass of silica plus water, in which fine clouds of gas bubbles rose uniformly. These clouds formed layered zones in the soft mass. As the total mass began to crystallize, the gas bubble clouds were trapped in the crystals, showing that the many and variously sized crystals of such a cavern emerged from *one* uniform soft *groundmass* that filled the entire space.

The old idea that rock crystal, like salt, crystallizes from an aqueous solution is thus obsolete. There are many indications that all crystallization connected with silica and the ores of metals have originated from such a gelatinous state.

But this gelatinous state or gel is that state which is of fundamental importance to living matter. All protein of plants and animals is in this gelatinous colloidal state and is permeated with living water. Through this living fluidity, the substance can become permeated by those formative forces acting from the periphery and the earth's interior. The gelatinous mass of silica in the rock cavities was once bound to the earth's life forces. When the "life" withdrew from the solidifying rock, it left traces in the multiplicity of crystal forms and the rocks surrounding them.

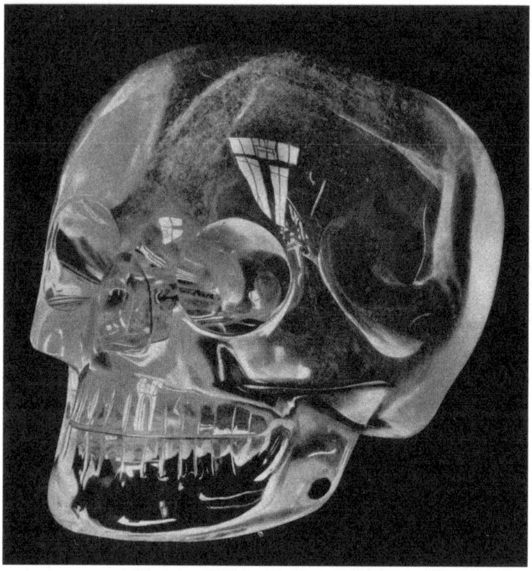

Figure 4: Rock Crystal Skull, British Museum, London.

## 3. The Siblings of the Rock Crystal: Smoky Quartz, Amethyst, and Citrine

In many places known for their beautiful, well-developed rock crystals, you can also find a noticeably darker colored variety of this gemstone known as smoky quartz. The hues of these rock crystals range from a subtle reddish-gray to clove brown and smoky gray to deep black. The latter is almost an opaque, black crystal, also called morion. Like rock crystals, these varieties were formed at the same time in the earth's history; they are relatively young compared to other gemstones. Therefore, the crystallization laws are also the same as we have understood them with the rock crystal.

Figure 5: Amethyst specimen from Magaliesburg, South Africa.
(Photo JJ Harrison, Courtesy of Wikimedia Commons)[iv]

Something is revealed in these smoky quartz crystals and in amethysts and citrines that we will study more extensively later in the "older" gemstones – ruby, sapphire, tourmaline, and others – namely *the secret of gemstone coloration*. This coloration of gemstones is based on the distribution of extraordinarily fine substances. One can call it a colloidal distribution in the same sense as shown in our explanations about the rock crystal regarding silica. The coloration of smoky quartz, amethyst, and citrine is based on such finely distributed substances. The remarkable thing is that the variety of colors we know from most gemstones does not appear in the rock crystal. There are no blue, red, or green rock crystals, only smoky quartz, amethyst, and citrine. The amethyst occupies a special position that we will discuss later.

The fact that the rock crystal has so few color varieties suggests that the substances that gave the other gemstones their colors were no longer present in the colloidal distribution at the time of its formation. There is quite interesting evidence for this thought from rock crystals that have been found enclosing other crystals. Best known is the rutilated quartz, also known as the Venus hairstone, which are rock crystals enclosing colored transparent minerals. The enclosed minerals are mostly hair crystals crisscrossed in a clear crystallized mass, which shows they had already solidified *before* the rock crystal. Such minerals as green actinolite, green and black hornblende, yellow to red rutile, needle-shaped manganese ore, and needle-iron ore (goethite) are embedded in rock crystals. The coloring substances of these embedded crystals are mainly iron and manganese. The author has had the opportunity to see rock crystals containing embedded golden-brown precious topaz and well-developed dodecahedron-shaped iron pyrites.

These phenomena indicate that the rock crystal is a younger form enclosing the older, already solidified crystals. However, these older crystals are mostly in some way colored, either as ore (needle ironstone, manganese) or as a mineral colored by a metal (actinolite, hornblende, rutile, topaz, tourmaline). It is significant that the same metals that give other gemstones their color, namely iron, manganese, and titanium (a rarer metal related to iron), have also been found in smoky quartz, citrine, and amethyst. Traces of titanium have been found in smoky quartz, where manganese and iron clearly do not play a role. The titanium is so finely distributed in the smoky quartz and is present in such a form that when the crystal is heated, the smoky color disappears, the crystal becomes colorless, and remains colorless after cooling. Let's now examine other gemstones, which also contain titanium as a coloring substance, for example, the sapphire. In this case, we see that the blue color *cannot* be destroyed by heating. After cooling down, the stone, which has been decolorized by heat, becomes blue again. This indicates that titanium, as a coloring metal, must be present in a different form in sapphire than in smoky quartz. The gray color of the smoky quartz shows that titanium must be in a *coarser* form leading to tones of almost opaque black (morion). Thus, we are confronted with the phenomenon where a substance in *older* gemstones such as sapphire is so finely distributed that it can produce blue tones. And in more recent forms like the smoky quartz, it appears in such a "coarse" form that it can only be active in darkening without actual coloring. *With the phenomena of gemstone coloration, we catch the tip of a process of fundamental importance for the history of the earth's substance.* It turns out that the metals in earlier, more living states of the earth were in a completely different condition than we know them today in the solid, opaque ores and minerals.

Rudolf Steiner has shown that metals were present as colored cloud formations in the "atmosphere" at a certain stage of the earth's development. This "atmosphere" must be imagined as a fluidic-air element that was still strongly permeated with the effects of warmth. One can say that the metals were colloidally dissolved in this warmed air-water atmosphere, similar to what we can produce in the laboratory today. When metals or metal oxides are brought into solution or fine distribution by certain chemical or physical processes, colored, transparent liquids can be obtained corresponding to the colors of the gemstones. Also, by evaporating the metals at very high temperatures in a vacuum and depositing the metal vapor on cold surfaces of quartz or glass, the most brilliant rainbow colors of colloidal metal layers can be produced. Such colloidal metal coatings are now widely used in the optical industry to improve optical glass in photographic apparatus and sunglasses. If you look around and observe, you can daily admire the splendid colloidal colors of metals.

The smoky quartz is the gemstone that picked up the titanium metal in a finely dispersed state at the last moment, so to speak. The titanium was already so concentrated that it could no longer have a coloring effect but only a clouding, darkening effect. A little "later," the titanium appears as fine needles of yellow to red rutile (titanium oxide) embedded in the clear, colorless, or slightly gray rock crystal in the rutilated quartz described above. Then the densification of titanium to the mineral stage came to completion.

The localities for finding good smoky quartz are extraordinarily widespread. The most famous and abundant smoky quartz of lighter to a deeply dark coloration was found in weathered granite in a crystal cave at Tiefen Glacier in the canton of Uri (Switzerland) in August 1868. This cave was 6 meters long by 4 meters wide by 1-2 meters high and yielded about 15,000 kilograms of crystals. The most beautiful of these crystals are exhibited in the Bern Natural History Museum and have special names. The "Grandfather" is 69 centimeters long and 122 centimeters in circumference; it weighs 133.5 kilograms; the "King" is 87 centimeters long, 100 centimeters in circumference, and weighs 127.5 kilograms. Among these huge crystals is also a doubly terminated one 82 centimeters long and 71 centimeters in circumference, weighing 67 kilograms. (See the description of the doubly terminated crystals in chapter 2. The Rock Crystal).

Smoky quartz is a companion to gemstones in the well-known mines of Mursinsk and Shaitanka in the Urals. Beautiful crystals are also found at Hinojosa del Duque in the province of Cordoba, Spain. In Madagascar and Brazil, the smoky quartz is found with rock crystals. Smoky quartz from Madagascar is considered the best. In earlier times, the so-called Scottish topaz found at the mountain Cairngorm in northern Scotland was important. It was recovered in the weathered deposits and in the riverbeds in 25 to 40 pounds weight specimens as clear transparent rounded pieces. Particularly numerous are smoky quartz sites on the North American continent, where it occurs exclusively in the granites.

Figure 6: An artistic rendering of Smokey Quartz

*Amethyst* differs from rock crystal and smokey quartz not only through its striking violet coloration but also through its somewhat modified crystal form. This difference becomes particularly clear where the amethyst lines the hollow spaces of agate geodes. The column, the actual prism of the rock crystal, is not developed at all, only the tip; the crystal continues down into a fused mass. Thus it happens that amethysts in geodes sit tip to tip like the teeth in a predator's jaw. The most intense coloration of the crystal is at the tip. When you look at an amethyst druse, you get the feeling that the color has been blown into the crystal from above. This only-in-the-top coloration is even more evident in other amethysts, which is characteristic for certain gneisses, sandstones, and pegmatite dikes (pegmatite is a granitic dike with long, lenticular cavities filled with crystals). Amethysts in these rocks have a column formed like the rock crystal, but the column is colorless or only lightly colored, with the main coloration sitting at the top of the crystal. Among these crystals, it is possible to find the so-called scepter crystals characterized by a thin colorless column and a thick head of amethyst on the top.

This gemstone exhibits another property that rarely appears in other gemstones. The coloring of the amethyst is not uniform throughout or cloudy but is embedded in the stone with fine stratifications (lamellae). If one examines the stone with a magnifying glass or a microscope, it is possible to find up to 30 finely-colored micro-stratifications between the layers and the twinning of the crystal. Between the colored layers, 1-2 millimeter thick lamellae appear, which can be colorless. This phenomenon is reminiscent of the layering of agates. Only here, in the case of the amethyst, which can appear in the interior of such agate geodes, it is not oriented spherically but is formed in strict accordance with the laws of a crystal structure. It is clear from this phenomenon that amethyst crystals formed rhythmically. In a certain way, this phenomenon reminds one of the remarkable structure of the phantom quartz crystals described with the rock crystal. The fine gas bubbles enclosed during solidification from the fluid state can be found with appropriate magnification in these lamellas. If you recall those remarkable gas inclusions observed in the rock crystal as horizontal layers, then it becomes clear that the amethyst crystallized in a completely different way than the rock crystal or the smoky quartz.

Figure 7: Phantom quartz crystal, Blue Phantom Mine, Arkansas.
(Photo by Rob Lavinsky, Courtesy of Wikimedia Commons)[v]

Adding the spiral tendency of the twinning formation of "right-handed and left-handed quartz" – a tendency that the amethyst also exhibits – to these manifestations of lamellar structures, then two phenomena can be seen in these crystal formations that are otherwise only known in the rhythmic and spiral formation tendencies of plant growth. It becomes clear from such observations of the fine structure of the crystals that even within the strict laws of crystallography, something still shines through of the primal relationship between silica (quartz) and the plant world, to which Rudolf Steiner has repeatedly referred.

The lamellar structure of these crystals and the type and intensity of the coloration are so characteristic of the amethyst that an expert can recognize where the amethyst originated. The amethyst has a kind of "inner habit" that allows an expert to recognize its place of origin by its fine inner structure; in the case of rock crystal, it was the outer habit that revealed its place of origin.

From the most delicate violet to almost purple (rare), the color range of amethyst is caused by a distribution of finely dispersed colloidal manganese, iron, and titanium. The amounts involved are from 0.1 to 0.001 percent. The substances are often only detectable with a spectroscope. In addition, nitrogen has been found in amethyst, suggesting that the coloration is due to a rhodanese-iron compound (ferric thiocyanate is a sulfur-nitrogen compound of iron). It is not so far-fetched to assume this organic chemical compound (rhodanese enzymes occur in the living realms of nature) is the coloring substance. There are rock crystals and amethysts that contain liquid carbonic acid, petrol-like liquids, or even hydrogen sulfide enclosed in small bubbles. From such phenomena, it can be seen that the atmosphere at the time when the crystals formed had a completely different composition than today.

For a long time, the best amethysts have come from Brazil and Uruguay. They come from the agate geodes mentioned above and are embedded in a curious rock formation, the melaphyre[4] (melaphyre-). This stone is grayish-green to black and fine-grained and comes from a period of the earth's history immediately following the bituminous coal forming period. The melaphyre

---

[4] Melaphyr-Mandelstein is an old name in German usage and otherwise not in common usage. See the photo "Achate im Melaphyr" at https://www.mineralienatlas.de/lexikon/index.php/RockData?lang=en&rock=Melaphyr-Mandelstein.

(actually black porphyry) is today considered a volcanic rock. However, it is important to consider that the "volcanism" of that time lacked the fire-breathing character of today's volcanoes. The rock masses forming the earth were still mushy and molten and not in a glowing fluidic condition. This soft mass was warmed through, and this warmth was not dead warmth but the expression of "life processes." The essence of the life processes was plant-animal-like and had not yet produced today's life forms; at that time, they expressed themselves in the still undifferentiated shapes of agate nodules and geodes.

These agate forms are colloidal and fibrous silica, exuded in fine, annular ring-like stratifications in the cavities of the melaphyre. This silica is always somewhat hydrous and colored in individual layers by finely distributed metallic inclusions (iron and manganese). A smaller or larger cavity in many such agate geodes remained during their formation and is lined inside with quartz crystals or amethysts. The color of the agate, however, was not transferred to the crystals forming inside. As a result, such geodes that contain amethysts can be in various varieties of gray or brown. Agate itself is never violet-colored like amethyst.

When Rudolf Steiner says that something plant- and animal-like found their last expression in these porphyries, one can point out that something "plant-like" is actually present in the silica and the remarkable annual-ring-like structures. On the other hand, the cavity formation of the agate geodes points to something "animal-like." Later, when we look at the agates, several other phenomena show us that these remarkable formations owe their existence to "organic" activity.

The elongated cavities can have different sizes. Those 1 meter long and 50-60 centimeters in circumference in an egg form are not uncommon. The largest geode excavated in 1900 north of Santa Cruz in Brazil had a dimension of 10 x 5 x 3 meters and contained 36,000 kilograms of fist-sized amethyst crystals. Other occurrences of amethyst are found embedded in sandstones and pegmatites in Brazil and Uruguay. The crystals do not sit in spherical cavities but line the walls of long fissures and clefts. The amethysts of Uruguay are characterized by a special depth of color. For such dark stones, Madagascar must be mentioned. Siberian amethysts are special, and those from Mursinka in the Urals are considered the finest amethysts of all. These Siberian stones are lighter in color, but they exhibit a beautiful blue color and a strong fire. They can also be found in many North American locations, but these stones do not achieve the quality of those from South America. It should also be mentioned that stones containing particularly large fluid inclusions are found in Georgia (USA).

Especially beautiful amethysts are found in Sri Lanka together with other gemstones coming from there. They come from granites and gneisses and are extracted from mountain rubble. In past times, the amethyst of Idar-Oberstein on the Nahe River in Germany's Hunsrück region was significant for the area. Like in Brazil, it was found in the elongated cavities of melaphyre. The few amethysts occasionally found in the Alps (Zillertal) are similar in color to the Siberian ones and are found in gneiss.

The *citrine* is a yellow-colored rock crystal whose color range lies between a light wine yellow to golden yellow. It is mainly found together with rock crystal in some places and in the company of amethyst in others. Its crystals become larger than those of the amethyst but never reach the giant size of a rock crystal. The remarkable fact that many amethysts can be made yellow by heating them to certain temperatures has led to the assumption that amethyst was formed at lower temperatures (100-125 degrees Celcius) and natural citrine at higher temperatures. However, the change to yellow in heated amethysts occurs only at 600-700 degrees.

Figure 8: Citrine, Russia.
(Photo Courtesy of Wikimedia Commons)[vi]

Two facts indicate that temperature alone cannot be the determining factor for the coloring of the citrine. The first is that citrine is never found in agate geodes like amethyst. It may accompany the amethyst when found in gneiss and granite, although it never shows the fine lamellar color layering that is so characteristic for amethyst, as described above. In addition, during the artificial firing process, the color of all amethysts becomes substantially darker than natural citrines. The expert can without difficulty distinguish a fired amethyst that has turned yellow from a natural yellow citrine. The coloration of the natural citrine is often cloudy but never finely lamellar, as with the amethyst. The coloration of the yellow-fired amethysts is still distributed by the fine layers described for the amethyst. Because it is possible to fire the more numerous amethysts to yellow, only a very few genuine citrines are left in the trade. Most of what is offered as so-called topaz, like the Spanish, Madeira, or gold topaz, is nothing other than these fired amethysts, whose hue is decisive for their naming, even if the amethyst used to make it comes from Brazil. The citrine's main sources are Brazil, Madagascar, Spain, the

Mursinka mines in the Urals, Colorado, North Carolina, Hungary, and Croatia.

Of the three siblings of the rock crystal – smoky quartz, amethyst, and citrine – amethyst is the one that has enjoyed a special reputation as a gemstone since ancient times. Its deep violet color spreads an earnest and solemn mood of humility and piety. Yet, at the same time, alertness and clarity are carried by this violet hue's calmness and modesty. Findings in ancient Aztec tombs indicate that this gemstone was once used for cultic purposes. Today, it still displays a lasting trace of its former dignity as a sacred stone in the bishop's ring.

Whoever deals with the amethyst in the present can learn something of the last act of those cataclysmic processes that took place when the earth began to release the last traces of life into its rock world. From the already darkening glow of the color clouds of the metals that once surrounded the earth, its violet descended into this stone and announced the times of awakening human consciousness. After that, the old colorfulness of the world faded to the yellow of citrine, the icy clarity of the rock crystal, and to a shadow in the smoky quartz.

## 4. About the Mothers of the Rock Crystal: Agate, Onyx, Jasper, Carnelian, and Tiger's Eye

There was a time in the earth's long history when nothing was created in solid forms, as is the case today. There were no plants that blossomed and withered, there were no animals that ran or flew about, there were no solid rocks, nor was there yet the human being who could walk over this earth and enjoy the forms of nature, its life, and its colors. Rudolf Steiner describes this time in his book "An Outline of Occult Science" and in the lecture cycle "Mystery Centers," which he gave in Dornach at the end of 1923.

It is not an abstract representation of the life of the earth at that time but a mighty epic about the unfolding of the realms of nature by the creative work of a hierarchical and cooperative circle of high spiritual beings. What lives in the Germanic creation myths (the Edda) and in the legends of world creation of all cultures of the earth as a huge picture of the earth's past, is in Rudolf Steiner's descriptions newly-created into a tableau of the world's origin, which can convey the same ancient truths hidden in the myths and legends of prehistoric times to the human consciousness of *today*.

A characteristic of all these old myths is that the world of rocks and living nature comes about from something originally *alive*, a giant, or a world-sized being killed or cut into pieces. This description corresponds to Rudolf Steiner's description that the whole earth was throughout a living being, which in certain times unfolded its life more plant-like, in other times more animal-plant-like, or only animal-like. It is then described how the silica coming into this living earth substance was the beginning of plant formation and how lime formed animal shapes through its activity. Silica and lime are the tools of the divine beings mentioned above.

The living substance of those ancient times was something that can, to a certain extent, be compared with the fluid content of a bird's egg penetrated by warmth, air, and light, only even more finely colloidal. When the silica began to work into this primordial albumen, the first plant forms appeared inside as if floating. These oldest plant forms were of a primitive kind, algae, and moss-like forms. They had the peculiarity that they had nothing permanent about them but appeared for a short time green, shining and fluctuating, only to disappear again, thereby making room for new forms to appear. It was like a rhythmic surging of plant life, of which a picture can be experienced when you have the opportunity to observe the back-and-forth movement of surging kelp forests in aquariums or under the sea. Indeed the algae, kelp, mosses, and some simple ferns are something like memories of earth's life during those most ancient times. However, one must never forget that these first plant forms had no permanency but repeatedly disappeared and emerged anew.

This weaving of the plant-like forms into the living "albumen" is closely related to the *activity* of silica. The silica that at first was something like a scaffolding for these plant forms, although still soft and waxy, later falls away and becomes the beginning of the oldest rocks of the earth, the granites, gneisses, and schists. What falls away is still soft and plastic; it is gel-like, dries up, and solidifies much later into the solid and hard rock we know.

Rudolf Steiner even points out in his lectures about the "Mystery Centers" that traces of ancient life on earth can still be found in today's rocks. He says: "Anyone who has looked around a little at nature knows quite well that something like distinctive marks of an ancient time can be found in the mineral world today. We find stones, we take them in our hands, we look attentively at them, and we find they have within them something like the outline of a plant form."[5]

Here Rudolf Steiner likely means the so-called *moss agates* found in India, North America, and other places. These formations consist of a milky translucent siliceous mass interspersed with a green substance shaped like algae or tree moss. This milky, translucent, often bluish siliceous mass consists of *chalcedony*, a form of silica characterized by its extremely fine fibers, although it has a habit like a gel. However, the algae or

---

[5] Rudolf Steiner, *Mystery Centers*, Lecture V, December 1, 1923, Dornach.

moss-like inclusions in this chalcedony mass consist of a completely different substance, namely grain-like hornblende-asbestos or the related minerals chlorite, celadonite, or glauconite. All these green minerals are siliceous compounds of iron together with alumina, potash, and magnesium oxide. These plant-like inclusions are very characteristic because they contain magnesium, which plays a decisive role in forming chlorophyll in plants.

It must be emphasized here that we do not believe these plant-like inclusions in the moss agate are plant traces as we know them from more recent layers in the earth. It is much more like traces of a "life," different in form and substance, that could hold in a living state what later became mineral. If one considers that today very old plant forms exist, like the so-called clubmosses (Selaginella) and our native ferns; also, the strange Proteaceae of Australia, which forms abundant alumina (aluminum oxide) in their mineral-nutrient budget and thereby differs significantly from the higher plants, then it becomes understandable that completely different minerals could be formed from plant-life that has now been lost.

It is important to realize that the green minerals enclosed in the moss agate's chalcedony groundmass can also be found as independent minerals. This is mainly the case with *asbestos*, which consists essentially of a fibrous magnesium silicate mineral. Its structure makes it suitable for weaving fireproof materials. Moreover, it exhibits a property that is otherwise only found in plant and animal fibers. Strictly speaking, the structure of asbestos is thus an organic one, even plant-like. Thus, we have before us a mineral that clearly shows that an organic-forming principle has overtaken the mineral-crystallizing principle. Therefore, the asbestos must also be regarded as traces of those ancient mineral-plant life-forms of which Rudolf Steiner speaks.

Closely related to moss agates and occurring together with them are those gemstones described as *bloodstone* and *plasma*. They are formed when algae or tree-shaped asbestos suffuses the chalcedony matrix causing it to become an opaque green. The plasma is formed in this way, while bright red spots are strewn in the rich green chalcedony groundmass in the bloodstone. These spots are said to consist of red iron ocher. Thus, this stone, which is also called *blood jasper*, is like some other gemstones in that it contains iron in two quite different forms, whose common origin can hardly be deduced from the laws of inorganic chemistry.

The rocks in which the three referenced gemstones are found are so-called porphyries and basalt-like rocks, whose "volcanic" character points realistically to a long, soft-plastic state of a mineral substance that has fallen away from life.

Figure 9: Agate and Moss Agate

Through the descriptions above, we have entered the oldest life processes of the earth that are connected with the activity of silica at the beginning of plant life. This silica still does not appear to us as rock crystal, but almost exclusively in that "mother state" of gel, which we have written about in the case of the rock crystal and its "siblings." In this "mother state," silica has preserved something of the plant-like origin of the earth's history in innumerable minerals, in precious and semi-precious gemstones. We could find those phenomena in rock crystal and its siblings as spiraling and twinning tendencies, phenomena that point to rhythmic, plant-like processes. We see these older formations of silica again in a different form. There is the fibrous chalcedony that we find in moss agate and the magnificent, pure blue chalcedony that has no inclusions. This colorful and varied mother substance of the rock crystal is embedded in kidney-shaped (reniform) and rounded forms, veins, and nests embedded in the rock, or it accompanies the ores that are brightly colored by the metals in all colors of the rainbow. Of these, only the aforementioned blue chalcedony and its green variety called *chrysoprase* are well known. This chrysoprase is a chalcedony colored apple green by nickel silicate. It is found almost exclusively in nickel deposits, and the only significant occurrence is at Koźmice, Silesia (Poland). These chalcedonies all have an exceedingly fine fibrous woody structure. One has the impression that the gel mass from which they emerged retained its softness and plasticity for a very long time before it solidified. During construction of the Simplon Tunnel, such silica gel masses that had not solidified or dried out were found at a great depth under the rock mass. It can also occur as a rarity in agate geodes.

This fine fibrousness also appears in the rhythmic layers and courses of the *agates*, where the layers of fibrous chalcedony alternate with completely colloidal, hydrous opal. We see how two states of the silica, the fibrous chalcedony and the still quite structureless opal, rhythmically interpenetrate each other in a fineness that only the microscope reveals: 6,000 to 7,000 alternating layers have been found per centimeter in such agates, always a chalcedony layer, then an opal layer, again a chalcedony layer, an opal layer, *600 to 700 layers to the millimeter*. Sometimes variously colored chalcedony layers follow each other in this way, but it is always a rhythmic process developed in unimaginable fineness, as in a living organism.

Where do we find such fine "oscillations" in the processes of life? Where do oscillations that are not solidified like the agate pulsate continuously in the life process and renew themselves? We find them, for example, in the cell growth of plants, in the formation of wood in phenomena such as annual tree rings. But we also find them quite freely and unsolidified in the fluid of the cochlea in the ear, where sounds of the outer world cause very fine vibrations in the fluid.

Figure 10: Onyx agate
(Photo Courtesy of Wikimedia Commons)[vii]

If we say that the onyx, with its white and brown-black layers, also belongs to the agates and has an inner developmental relationship to the *sense of hearing* (see chapter 2. The Rock Crystal); it is then possible to understand that in the rhythmic layers of the onyx-agates, something like a mineral image of the process of hearing is present. Closely related to the onyx is the *carnelian*, which is found together with the agates but is distinguished from them because, in its purest and noblest state, it is completely free of layers and courses, has a deep, dark red color, and is translucent. It is also a chalcedony colored red by iron with finely-fibered silica as its base. Real, natural, undyed red carnelians come exclusively from India. Today, most of what is on the market are layerless agates colored by iron solutions and then fired; they are frequently found in Brazil. The same is true of onyx, which has become extremely rare in its brown-black-white alternation of layers. A version is produced artificially by soaking gray-white agate in honey water and then placing the stones in concentrated sulfuric acid, burning the honey's sugar to black charcoal.

Onyx and carnelian are found in the remarkable cavities of melaphyre-mandelstein[6] and are connected with the last phases of the plant-animal life stage of the earth. We have here the most outstanding occurrence of agate fillings in cavities that are interspersed like an organ. If one wants to compare this form with the animal world, think of the gastrula forms (hollow and pocket forms) of lower animals. The "life" underlying the agate formation that is reflected in the extremely fine layering was a plant-animal: the "animal" points to a hollow space, the "plant" to the silica that gives it its substance. This substance is not structured like a mineral but is formed organically. The filling of the cavities with silica proceeded from the outside to the inside, and at the center of the inside, rock crystal or amethyst formed when a cavity remained to facilitate their crystallization. According to new research, the layered formation of the silica substance was already present *before* differentiation into fibrous chalcedony and colloidal opal.

A most significant phenomenon in agate is the *spherical* deposition of its layers, which obviously must have proceeded *independently of gravity*. Each agate nodule has a so-called injection hole through which the siliceous substance entered the cavity. This hole leads through all layers to the innermost space of the cavity.

---

[6] Melaphyre-mandelstein (Melaphyr-mandelstein): In German usage, melaphyr is a fine-grained rock, sometimes with a porphyry structure. They often have large cavities that are partially or completely filled with secondary minerals, the so called "almond stones." It's a very old name replacing trapporphyr and originally used for amygdaloidal rocks composed of plagioclase and augite but later used for Upper Palaeozoic rocks, usually basaltic.

There are also agates where spherical layering did *not* occur; the layers are parallel and *horizontal* to the cavity's position in the stone. In this case, gravity obviously caused the positioning of the layers; the spherically orienting organic principle is no longer active. Such rocks are called *sardonyx* and are only found in Uruguay. Due to the leveled position of the layers, they are particularly suitable for gem cutting. The sardonyx should not be confused with the *sard*, which is nothing other than a dark brown chalcedony. In ancient times it was imported from India to Greece via Sardis to make jewelry.

Some other gemstones can also be considered to be mothers of the rock crystal. These minerals contain silica in a fibrous form similar to chalcedony but with a coarse texture visible to the naked eye. These stones are called the quartz *cat's eye*, the *tiger's eye*, and the *falcon's eye*, three precious stones characterized by fine-grained quartz closely intergrown with extremely fine hornblende asbestos fibers. In the case of quartz cat's eye, this asbestos is usually olive-green so that when the stone is cut, a rounded, raised form of cat's eye-like iridescence is visible when the stone is moved. The stone is opaque. The cat's eyes are found in Sri Lanka, East India, and the Fichtel Mountains in Bavaria, Germany. They are found in rocks that often contain asbestos. With the falcon's eye, the hornblende asbestos is dark blue-green. Its fibers reach a thickness of only a thousandth of a millimeter and are closely intergrown with the coarser stem-like fibrous quartz. The tiger's eye is formed from the falcon's eye when the iron content of the hornblende asbestos weathers and oxidizes, giving it a rusty yellow color. Like the cat's eye, both gemstones are intensely colored. They have a silky luster when correctly cut, which vividly shifts through the kinking and bending of the fibers.

When looking at the rough stones of the last three gems, one immediately has the impression of wood. Indeed this has to do with formations that belong to an earth age completely permeated with plant-like formative forces, which manifest here more like wood.

These gemstones are also mothers of the rock crystal so far as they reflect the primordial state and the silica's primordial activity as a formative plant force, giving us the possibility to create pictorial ideas about the beginnings of the earth's history.

To this large group of mothers that form the rock crystal belongs another gemstone that no longer shows much of the described living origins. It is the *jasper*. It is best known for its red color but differs from carnelian in that it is completely opaque, and only in some places can it be found with a fibrous structure. Its basic structure is fine-grained. The coloring substance is mainly iron. The actual color depends on the oxidation state and the metal distribution and can cause many yellow, brown, red, green, and even blue and black tones. In most cases, the colors are not uniform, but the stone is cloudy, spotted or striped, and colored by a predominant base tone.

The variety of these jasper formations is extraordinary. Like the agate, they are predominantly connected with "volcanic" rocks, whose actual nature we have already described. In rock, the mineral is found as layers, courses, irregular nodules, fissure fillings, and, again similar to agate, in spherical forms. The latter, the so-called ball jasper, is mostly red and is found in southern Breisgau near Auggen and Mulheim in Germany. A brown jasper of this type comes from Egypt (Nile silica), colored chestnut brown and brick-red. North America has magnificent brown, red, and yellow jasper deposits in stone quarries at Sioux Falls, South Dakota. A yellow jasper comes from Sicily and a blue one from Grossalmerode in Hesse and Nordböhmen.

The jasper substance comes from a time of earth formation when silica had already fallen far more out of the life processes than at the time of the agates and chalcedony. For this reason, we find hardly any fibrous structure. But the fine distribution of the metals that provides the color is still present. The silica is gelatinous and plastic but is no longer formed by the disappearing plant forces. Thus, only a finely granular, crystalline structure is formed. However, the silica must also have had a much greater solubility in water at that time because it shows itself to us now through the fossilization of organic forms. This fact becomes obvious with *petrified wood*. Even though it is not an actual gemstone, it often shows in its magical coloration something of the essence of a real gemstone. Best-known is the petrified wood found in Arizona in North America.

There is an area of actual petrified forest where some trunks are still standing upright; hundreds and thousands of powerful trees lie completely silicified on the ground in ash-like weathered earth. The diameters of the trunks are so large that one could cut and grind small stone tabletops from the cross-sections. It is difficult to imagine the patterning and colorfulness of these trunk cross-sections. Occasionally, it is possible to see the petrified woods of Arizona in collections and museums. Unfortunately, the entire area has been placed under protection after unimaginable overexploitation by enterprising stone seekers.

The occurrences of these petrified woods belong to two quite different earth ages, an older one, which follows the coal age (the Arizona woods belong to it), and a younger one before the large dinosaurs appeared. One sees from this that certain developments in the characteristics of the materials *repeat* themselves. Perhaps in the intervening period, the silica had quite different properties than at the time when the petrified woods could form. The trees from which these petrified woods were formed are similar to the South American araucaria, known in Europe as "evergreen conifers," except that the wood of these "antediluvian" trees was light and airy, perhaps like elderberry pulp. In these silica-permeated woods, the circle closes. It began with the arrival of plant forms through the silica and ends with the preserved plant form permeated with siliceous substance. What lies as silica outside this circle obeys other laws.

## 5. Rose Quartz and Opal

In the most hidden depths of the primary rocks of our earth rests a gemstone that has always delighted people with the uniqueness of its color: the *rose quartz*. In the relatively few places where it emerges from the mysterious pegmatite dikes in granite, it is usually present in such quantities that it is not in itself precious or rare. Whoever has had the opportunity to find rose quartz in nature, as it literally blossoms forth from the uniform colors of gray and yellowish rock, must ask: how is it that this color, which we otherwise know only from the delicate blossoms of roses or the shimmering skin of a child, appears here in the somber rock?

To answer such a question even vaguely, it is necessary to look at the history of the rocks of our earth with completely different eyes from those existing in recent centuries. In some respects, we have already attempted this in the preceding chapters, but it is always necessary to bring in new points of view which shed light on the inner evolutionary connections between the rock world and the living realms of nature.

In his lectures about "Mystery Centers," Rudolf Steiner has shown that the oldest life processes of the entire "earth" organism were of mineral-plant and plant-animal nature. Their remaining traces can be found in the remarkable woodlike or organic plant-like structures in certain rocks and minerals, but also in the extraordinary colorfulness of many minerals from this "time." The clear organic conclusion of this life epoch of the earth is bituminous coal and petroleum, which corresponds to the etheric oils of our higher plants.

The onset of plant-animal life processes strongly overlaps with the preceding mineral-plant life processes, which show their residual traces in the already mentioned agates and similar formations. In the hollow cavity formations of the melaphyre, which is the mother rock of the agates, the life processes could be said to move toward the animal. Such cavity formations also occur in the granites, thus forming the foundation of the earth's mineral layers in those very old rocks. These cavities are not spherical like agate geodes but elongated, tubular, and can extend for long distances. They meander through the granite, narrowing and widening again into large cavities. The components of granite – quartz, mica, and feldspar – become coarse-grained in huge masses along the margins of these tube-like passages, and the feldspar and mica become gigantic crystals. This metamorphosis of the granite's components to a granular coarseness, forming large crystal specimens with concurrent cavity formation, is called *pegmatite*. In these pegmatite cavities, the most beautiful quartz, mica, and feldspar crystals are formed and are accompanied by various colored and rare gemstones.

It is significant that the giant forms of the granite components do *not* sit as giant crystals in the hollow spaces but enclose the hollow spaces as coarse grains or tablets. Feldspar crystals can become large enough to host a complete quarry operation in a single tabular grain, which at the same time is a single crystal. Such a pegmatite with a large feldspar crystal was found in the Ural Mountains. In India, pegmatites are known to have mica books of more than one meter in width. In normal granite, these minerals have grain sizes of two to five millimeters! The *quartz* of many such pegmatites is transformed from its enormousness into *rose quartz*. Thus, while feldspar and mica change only their size but not their color, quartz, which in itself is gray or colorless, can appear in these pegmatites as rose quartz.

The enlargement of the components of granite around the pegmatite dikes, the appearance of gemstones in these dikes, and the sudden appearance of rose quartz indicate that completely different formative forces were at work around the veins than in the granite itself. Here something intervenes in the rock formation that we already characterized as an individualizing principle at the beginning of our considerations. This principle, which we have seen in countless metamorphoses in the living and spiritual realms, has also created an expression in mineral formations.

Looking in the human realm at the expression of a personality through its actions, gestures, and achievements, into the shape of its body and physiognomy, we know that in it something appears that lives and works in the present, lives to a high degree in consciousness, acts and forms itself and its environment.

We cannot look in the same way at the animal world. The individual animal is a member and expression of something above it that certainly creates form and being, but which does not directly express itself therein as a human being does. Therefore, spiritual science speaks about group souls of animals that are understood as high spiritual beings.

In the plant world, we only have to do with expressions of life that are not imbued with soul qualities. The soul element only hovers over the plant, which creates a weak reflection in the flower's essence. This essence that is creating and life-sustaining in the plant world must be sought for in different regions than the group soul of the animals and the human ego. It is working from the periphery of the earth, from the planetary spheres, and has its life in the change of the seasons.

In the mineral realm, we are dealing with that which has come to an end, been eliminated from the stream of life, which can appear to us as resting in a deep sleep. That which was actively creative has completely withdrawn from its work and left its creation to earthly laws, which destroy and repeatedly dissolve what was created. This *development* of the mineral realm is closely related to the development of the nature realms and human beings.

To understand this development of the mineral realm, it is important to consider the structure of the human head. The bony shell of lime enclosing the brain is here the last thing to be formed, just as in the mineral realm, the enormous lime formations of the Jurassic and Cretaceous cover the preceding layers of primary schistose rocks. Thus, in the layers of the brain, the schist's formative forces are reflected in a metamorphosed way; in the sense organs and their functions, we must look for those formative forces that created the so-called primary rocks in the mineral world. Only with this consideration does it become understandable why there is a developmental connection between the gemstones, which are found mainly in the primary rocks, and the sense organs and their functions.

The other aspect mentioned earlier concerns the origin of these primary rocks from mineral-plant-like life processes connected with the incoming silica. Rudolf Steiner characterized these earliest life processes in other contexts as a kind of blossoming, initially without any leaf or root functions. This sheds significant light from a different angle on the coloration of the minerals of the primary rocks. We already learned that this coloration is connected with finely distributed metals.

It is an important fact that metals play a great role in the coloration of the flowers of our present-day plants. Above all, iron and the related manganese (which we already learned about in the case of amethyst) are closely related either materially or functionally to the formation of flower pigments. But manganese is also the metal that gives rose quartz its unique *incarnate* (peach blossom) *color*. And the color of rose quartz is very sensitive to light. It is unpleasant for the gemstone lover to discover that many rose quartz specimens from certain locations lose their color in the light and become gray! One could almost speak of a flower-like sensitivity with this rose quartz color. Other gemstones do not exhibit this property at all.

Having pointed out the origin of the rose quartz from the old mineral-plant-like life processes of the earth, we can now see in this siliceous mineral that formed around the pegmatite cavities, something which is like an early image of that incarnate color we observe when iron-rich blood shimmers through the skin of the human being. This skin is an organ completely penetrated by silica activity and contains silica, enabling it to be a sense organ.

Consider the common rough quartz (not rock crystal) as that which in the rock world corresponds to skin formation in the human being. Then, we can take the rose quartz as a picture of that living process underlying the blood-suffused human skin. In this sense, the rose quartz is a mineral related to human health, that harmony of spirit, soul, and body expressed in the pure incarnate color of the blood-suffused skin.

The most beautiful rose quartz comes from Madagascar, that strange island in the Indian Ocean, which in its rocks, minerals, and gemstones occupies a remarkable place in the world. In the rose quartz from Madagascar, you can really see that this gemstone emerged from an ancient gel state, about which we have already spoken several times. In contrast to the appearance of rose quartz in other countries, Madagascar's rose quartz has richer color and is clearer, with many fewer flaws.

Significant locations of rose quartz are found in Brazil in the provinces of Minas Geraes and Bahia. Further places of discovery are in North America (Colorado, California, Maine, South Dakota), India, Japan, and South Africa. The deposits in Pisek, Czechia, and the Bavarian Forest at the Hühnerkobel near Bodenmais were already exploited in the Middle Ages and supplied the material for the gemstone walls in the St. Vitus Cathedral chapel in Prague and the Karlstein Chapel, which are mentioned in the first chapter. Rose quartz is also found in Scotland, the Hebrides, the Shetland Islands, Ireland, the Urals, Austria, and Transylvania in Romania. In all the places mentioned above, rose quartz is associated with the remarkable formation of pegmatite.

The *opal* is a gemstone that stands in contrast to the rose quartz. While the home of the rose quartz is found within the oldest foundations of the earth, the opal belongs to the youngest formations of the mineral world. These most colorful gemstones, which contain all the colors, are not based on finely dispersed metals but on a web of water and air that permeates the stone in gossamer structures. This may explain why the opal has been considered a stone with a special relationship to the lungs since ancient times. The lungs are the organ in which liquid and air touch each other intimately in the tiny pulmonary alveoli during the breathing process.

Figure 11: Opal from Yowah, Queensland, Australia
(Photo by Noodle Snacks, Courtesy of Wikimedia Commons)[viii]

The substance of the opal is hydrous silica, which has completely retained its original gel state and shows no trace of crystallization. The opal also differs from the crystalline rose quartz because the opal is rarely found in crystalline form. The water content of the opal can reach 34 percent, although it is about six to ten percent on average. The water can diminish as the stone dries out, causing the color play of the stone to diminish. However, in humid air, the stone can "breathe in" water again and regenerate itself. The ability of the opal substance to absorb water and release it again in dry air, changing its color play, is extraordinarily variable. The normal precious opal shouldn't dry out because its water content is decisive for the strength and variety of its color play. There are also opal varieties, such as the brown and opaque *hydrophanes* of Mexico, which are very porous and sometimes stick to the tongue. In the dry state, these Mexican opals exhibit a beautiful play of colors like the precious opal. However, if they are put into water, they become clear and transparent, and the play of colors disappears completely. With these stones, the air finely distributed within them seems to play a role in the emergence of the colors.

A true hydrophane is usually found together with precious opals and shows exactly the opposite behavior. The natural insignificantly colored, porous, and scarcely transparent stone becomes almost completely transparent when immersed in water and often shows a beautiful play of colors. Therefore this hydrophane opal has been called the eye of the world, lapis mutabilis, and chamaeleon stone. The same effect can be achieved by drenching the stone with oil, wax, or resin. These strange, most beautiful hydrophanes are found at Hubertusberg in Saxony, together with amethyst, rock crystal, and common opal.

The mineral is found in string cords and layers in the rock or enclosed with the minerals mentioned above in chalcedony spheres. In these bluish chalcedony spheres, the hydrophane is often found in an unhardened *gelatinous* state so that nature clearly shows here how an early state of the gemstone's mother substance can be preserved.

The contrasting behavior of the described hydrophanes reveals the primordial phenomena that become visible and constitute the essence of the opal, the aerated and hydrated silica substance from which light actively conjures up the colors. Light, air, and water are essential parts of the elements of life. The opal develops its mineral "life" in these elements, but we experience the completeness of its original life only when we look at the realm of the rock world in which it was formed. We are not led down into the primary rocks but into the very "recent" times of those processes, which we know today as *volcanism*. There we meet the fourth element of life, the *warmth* that has contributed to the formation of the opal. In this more "recent" time of volcanic

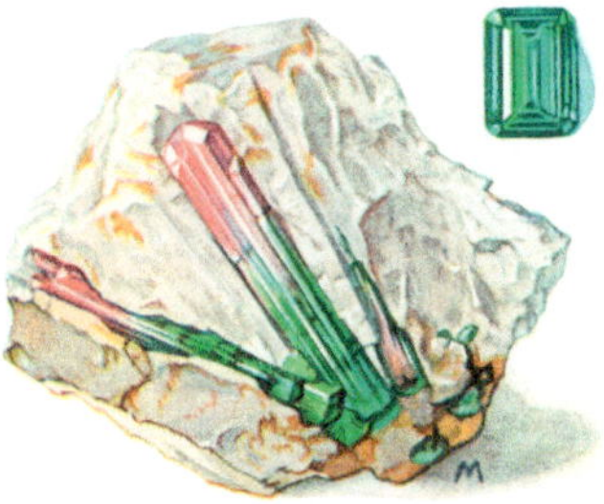

**Tourmaline Crystals**
**in Marble**
**Governor, New York**

**Chrysolite-olivine Crystals**
**in peridot**
**St. Johns Island, Red Sea**

Rock Crystal, Brazil

**Citrine Crystals, Madagascar**

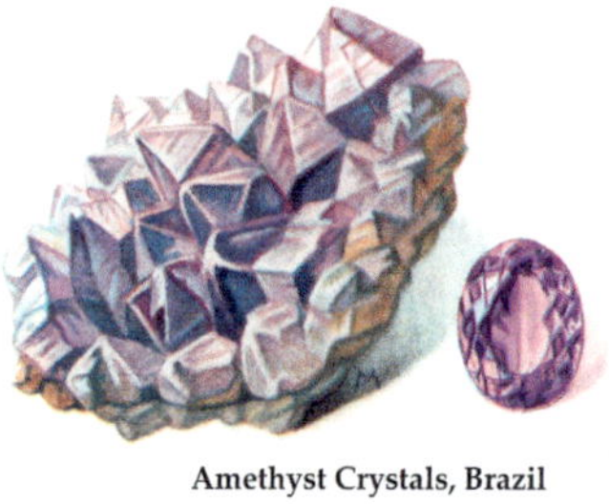

Amethyst Crystals, Brazil

**Rose Quartz Brazil**

Figure 12: Artistic representations of gemstones discussed in this book.

Figure 13: Artistic representations of gemstones discussed in this book.

processes, the entire rock world is still plastic and soft, but a part is not only in this state but also warmed through to a great extent. This thorough warming must not be imagined as a hot fluid heat but the evenly working together of water, air, and warmth, similar to a living organism. One could speak of hot mush, a mush that is not dead and unformed but permeated by forces that act on it from the outside. These forces also move it; they flow and stream into the cliffs and crevices of the earth's already cooler and firmer parts. It only solidifies much later and then becomes lifeless, and the forces are lost.

There is a whole series of phenomena in present-day volcanism that still show this state of warmed-through soft rock masses. The remarkable interaction of water, air, warmth, and solidity in the fluid magma in the interior of a volcano is revealed, for example, in the fact that this magma before the eruption in the depths of the earth has a temperature hundreds of degrees lower than after the eruption when it flows down the mountain as lava. In most cases, the molten magma inside the volcano has a temperature of less than 900 to 1000 degrees Celcius. However, it heats up rapidly in the air to 1100 to 1200 degrees Celsius when it erupts lava, emitting gases and steam, and then only solidifies very slowly. What is going on?

Considering that silica has a melting point of about 1800 degrees and that feldspar, mica, and other minerals containing silica require about 1500 degrees Celsius to become fluid, we can see the role of water and gases in the magma. Namely, when water and air are present, the silica does not need enormous heat to be fluid. Suppose one tries to imitate these inner conditions of the earth in the laboratory and melt the hardest rocks in the presence of water and gases. In that case, the mixture must be enclosed in steel containers because otherwise, the water and gases escape.  This process creates enormous pressure in these chambers, and the rock “melts” in the container at temperatures far below 1000 degrees.  In the earth's interior, the pressure is not as high as laboratory observations, which can be deceptive when such results are transferred to nature.

The fluidity of the magma in the volcano has nothing to do with pressure. Suppose probes are placed into the flowing magma in a crater. In that case, the magma, which flows openly in the air, has a lower temperature far below 1000 degrees a few meters under the flowing surface. But on the surface, where it touches the outside air, it is above 1000 degrees. This condition occurs because magma *decomposes* in the air. It falls apart into the air (gases), water vapor, and solids, but the warmth that held it all together becomes free and measurable. By adjusting the condition of today's volcanoes to a still lower temperature, the processes and conditions connected with opal formation are found.

Everywhere the opal is found, it is related to such ancient "volcanic" processes. Therefore, it is often linked to certain rocks, whose fissures and crevices are filled with kidney-shaped and grape-like (botryoidal) forms. The interweaving of opal substance with water and air is bound to a very fine structure, the essence of which was discovered early last century. It was found that the finely layered surfaces upon which the opalescence and interference phenomena of the opal color play depend have a lawfulness that corresponds mathematically to the inner structure of calcite crystals. Therefore, it was concluded that the opal substance had displaced calcite by its penetration into the fissures. Of this calcite, only the structure is found in the opal; calcium substance is not present. Whether or not this phenomenon observed in all precious opals is actually due to a "displacement" of calcium carbonate cannot be proven. Nor can one "prove" that the bee uses small rock crystals to form its combs accordingly. Nevertheless, the bee "knows" the form of the rock crystal. Similarly, the formative forces of calcium carbonate were "used" by the living earth to form the opal.

The opal appears differently at individual *places of discovery*. Australia produces opals from Queensland and New South Wales with the most beautiful red, green, and blue surfaces that interpenetrate at the edges. The opal substance often contains "fossilized" wood, shells, and even dinosaur bones. Some of the deposits yield the famous black opals in iridescent red, green, yellow, and purple. One of the largest rough black opals found there weighed nearly five kilograms. Other occurrences of Australia include the harlequin opal that exhibits colors in small round or angular patches. These opals are reddish-yellow to colorless in transmitted light; in reflected light, opalescence is seen on a milky-white to a grayish-blue or blackish background. The so-called Hungarian opals, which now come from the Czech area, are distinguished by a small-spotted play of violet-blue, blue-green, and red colors.

Mexico is the place to discover the beautiful fire opal, also called sun opal or girasol, with a brownish-yellow to brownish-red color and little opalescence. The fire opal is even more sensitive to light, air, water, and temperature than the other opals. Also found in Mexico is the so-called lechosos opal, which is fire red and exhibits an intense emerald green, crimson and dark violet-blue play of colors.

All opals in a mountain-moist condition are very sensitive to "drying out" too quickly. They may have to be stored for a long time in boxes or embedded in moist clay to avoid cracking and a loss of color. The seemingly dead stone turns out to be as "sensitive" as a flower plucked from the mother plant. Its home is the warm and humid soil of mother earth.

## 6. Precious Feldspar, Moonstone, Sunstone, Amazonite, and Labradorite

From the oldest life processes of the earth, which we characterized as mineral-plant-like in the preceding chapter about the rose quartz and the opal, a rock world separated that does not contain any traces or forms of life we would recognize from the life realms present today.

We are confronted with a great riddle when Rudolf Steiner describes the life processes of the earliest times as enormous flowering processes because we know of no life processes from which something similar to rock comes about. Of course, we know of processes where something is deposited from lush plant growth, which can become peat, brown coal, and bituminous coal, but these are not minerals, not rocks.

Only in much younger layers of the earth, for example, with lime, can we draw a comparison to processes we know from life. Every animal and human organism forms a calcareous bone structure that can be preserved for a long time after death. This bone structure can become mineral-like to a great extent because, in the deposition of phosphoric lime, rich crystals of this substance continue their existence independently of the bone-forming cells once the lime has been eliminated from them. From this example, we can see that a mineral substance can be separated from a living-organic process, although it can no longer take on an organic form. If sections of these human or animal bones are made so thin that light can shine through them, then bone cells can be seen with the microscope. They show themselves by a delicate polyhedral arrangement throughout the total mass. But in addition to this structure, there is another one, which does not coincide with the cell outlines; it is an angular shape, running crosswise through these cells and shows the exact crystal structure of the apatite that we know from this mineral, a calcareous phosphoric acid from the rock world.

*Looking at this remarkable world of the oldest silicious rocks and many later rocks as something that came about only by inorganic, chemical, and physical processes is making the same mistake as a researcher who merely looks at the bone through the microscope and claims it is only a mineral substance that has nothing to do with life. After death, the bone's cell structure disappears when the bone has lain in the ground for some time. It is then only possible to recognize it is a structure from life by its rough outer form.*

But let's look at the earth's enormous rock masses, the earth's bone system, as Rudolf Steiner called it again and again. Suppose we are in a very small section of a quarry, then we are in line with the researcher who believes the study of the bone's real essence can be done using the microscope alone.

Rudolf Steiner points out that when looking at plants under the microscope, it is possible to see how their substance, their essence, tends to be "rock-like." He meant by this that the cell structure of a plant mass looks under the microscope just as "granular" as the "granularity" of rock (granite or sandstone). This hint is extremely important because it shows the importance Rudolf Steiner gives to the *structures* of a substance. In nature, very different substances can have identical or similar structures, which speak of a similar underlying principle in substance formation. For example, just think about the boundary layer of animal or human cells consisting of protein, while plant cells have a carbohydrate covering, meaning a wood-like substance. The "cell-forming force" thus uses protein one time and woody substance another.

From this perspective, it is not so far-fetched to regard the cellular structure of many rocks, especially the silicious ones, as residual traces of a life activity belonging to the past. As we mentioned initially, this life activity was flower-like. Rudolf Steiner reported from his research that what we find today in the silicious primary rocks as mica was something similar to the formation we know in today's plants as sepals. On the other hand, *feldspar* is regarded as the solidified remnant of the style or pistil of these old mineral-plant flowers.

This long contemplation of the earth's past was necessary to better understand that the feldspars, as ingredients of granite, gneiss, and some other ancient rocks, are the carriers of the *color* of these rocks. Red, yellow, blue, and green colors, especially in the granites, come from colored feldspar in a mixture of quartz, mica, and feldspar. Thus, the colored element is most related to the inner flower parts, whereas the mica as a leaf form is almost colorless and appears in brownish to greenish to black tones. The colors of the feldspar come from finely distributed iron or other metals related to iron, for example, nickel. It is a similar phenomenon seen in many other gemstones we have already met.

Figure 14: Moonstone, Oslo, Norway.
(Photo by Lech Darski, Courtesy of Wikimedia Commons)[ix]

In addition to this coloration of the feldspars by metals, another type of coloration occurs, similar to the color phenomenon of opals. It is seen most purely in the *moonstone* (opalescent microcline). The fairy-like blue shimmer and the cat-eye effect (chatoyancy) of the unique stones of Ambalangoda in Sri Lanka are incomparable to any material from other sites. When a flat, roundish cut stone is placed on a black surface, a blue glow appears in the center of the stone, while a milky-silky and translucent zone surrounds the glow. Elliptical stones make the glow glide back and forth along the long axis when moved. None of this shimmer is visible in transmitted light.

The moonstone is, in essence, a so-called potassium feldspar, an orthoclase, a potassium aluminum silicate. In these feldspar gemstones, we meet for the first time in this work with silica *compounds*. Everything we have observed before has been exclusively the "pure" silica in its various colors and structures.

These feldspars are characteristic in that they are not consistently pure substances but can be, for example, mixtures such as potassium feldspar, sodium feldspar, and calcium feldspar. These natural mixtures are variable, and all have scientific names and formal definitions. But this shall not concern us.

The potassium feldspars predominate in the older feldspar-bearing rocks (granites, gneisses), namely those with a predominant potassium content. However, younger rocks that formed later also contain sodium-potassium feldspars and sodium-calcium feldspars containing practically no potassium. This change in feldspars is accompanied by a change in the other rock components. Mica, for example, is replaced by black or green hornblende, or the white potash of the older rocks becomes dark and iron-rich. The variety of these transformations is extraordinary and results in a truly flowery array of rock colors.

The blue shimmer of the moonstone can be physically and structurally seen because the sodium feldspar has "separated" in fine lamellae from the potassium feldspar. In a certain way, it has separated from the close and uniform interpenetration with the potassium feldspar and has brought its own somewhat different crystal structure to expression in the finest form. The finer these layers are, the more intense is the blue light of the moonstone. If the stones are heated, this separation is eliminated, and the blue shimmer disappears. This phenomenon indicates the structure cannot have originated from a firey flux, as is the current theory, because it is precisely the heating that eliminates the separation. One will certainly get a better idea of the moonstone's formation by looking at the pearl luster of a shell. There, from an organic separation process, the calcite and a fine, horn-like substance are secreted by the mollusk in the thinnest plates and layers on which the mother of pearl luster then appears through the diffraction and reflection of light.

Similar mother of pearl light effects can also be observed on the flower petals of some plants, which are caused by layered cells with a different filling. These formations in plants and animals all originate from gel-like states of substance. Considering that the gel state precedes crystal formation in the mineral realm, it is not far-fetched to presume the "separation" of the moonstone's components is a process that occurs primarily during the gel's transition to a crystalline form. High temperatures can not play a role in this process. The moonstone is feldspar and thus a granite component and is also one of the minerals found in the pegmatite dikes described for the rose quartz. However, the places where they are found are completely different; the moonstone is never found with the rose quartz.

As already mentioned, the most beautiful moonstones are found in Sri Lanka, where they are partly hidden in weathered masses of rock under the cover of thick tropical vegetation. Lesser qualities of the stones only show a bluish-white or pure-white glow. In the past moonstone-like adularia feldspar came from Switzerland, from the area around St. Gotthard. Other places where lesser qualities are found include North America, Brazil, and Australia. In Myanmar, it is possible to find some large moonstones together with rubies. These moonstones, however, exhibit only a white glow.

In the series of other feldspar types mentioned above, some also exhibit the moonstone effect. Albite, plagioclase, and oligoclase varieties of moonstone are found at various North American locations.

With the so-called labrador moonstone and its wonderful blue glow, we come across another type of feldspar, the *labradorite*. The labradorite is a sodium-calcium feldspar, which plays a major role in many of the earth's massive rock deposits. On the one hand, it can be a component of deep rocks like gabbro or porphyry, and on the other hand, it can form enormous rock masses as the so-called Labrador rock. The latter is interspersed with the finest quantities of magnetic iron ore or ilmenite (an iron-titanium mineral), which appears on the rock's polished surfaces as shinny metallic spots. The labradorite, which can be used as a gemstone, comes exclusively from the rock of this mineral when an individual "grain" reaches a certain size.

The colors appearing in this gemstone belong to the most surprising and beautiful that we know of in the mineral realm. The incomparable shimmer of tropical butterfly wings and the unique iridescence of tarnishing steel can be seen on this gemstone with their full range of colors. The most common color is blue, in all its shades from cobalt blue to violet. But green shades between emerald green and yellowish-blue green also commonly occur. Rarer are yellow and red tones, ranging between purest lemon yellow, golden yellow, orange, copper-red, and tombac-brown. The colors always have a strong metallic luster and move across the surface when the cut stone is turned. The basic color of the stone is an inconspicuous smoky gray, ash gray, sometimes blueish gray. When looking at split or cut stones, one is always tempted to hold the stone up to the light to appreciate its transparency, which leads to disappointment since this stone is completely opaque. This is because the colors of the Labradorite have an incomparable transparency. The shimmer seems to come from the innermost depth of a hidden light source in the stone.

What is the “cause” of this play of colors in the labradorite? Its appearance is similar to that of the moonstone because the labradorite’s inner crystal structure is such that light must pass through fine laminations of dense, opaque matter. It is simple for modern scientific thinking to say that the crystal structure that causes the iridescence to exist is an accidental one derived purely from the laws of crystal mechanics. Nothing has happened here but a chemical-physical process of a special kind, which has arranged matter by chance in such a way that when it is brought up from the depths of the earth, it shows these phenomena in the light.

Whosoever thinks and speaks in this way does the same as a researcher who would maintain that the dreamlike colors of the wings of the tropical butterfly owe their origin only to chemical-physical processes. It can be said with certainty that the material of these butterfly wings is a quite definite chemical substance with a certain structure, with which the light interacts so that the colors appear. But isn’t the real creator of these substances and their structure the essence of the living butterfly, which creates its colorful body from and with light? Aren’t the colors of the flowers and the butterflies in a special way dependent on their relation to the intensity of the sunlight over the entire earth?

We can see how the light blends with life processes and allows substances to reveal something of their essence, even in dense dead matter. If we look at labradorite or a moonstone, we must say to ourselves: *The creative powers of light have participated in the fine inner structures of these crystals*. In the times when the substance of the labradorite began to form, the earth must still have been completely suffused with light. This inner light of the earth formed the substance so that even today, it testifies to the living, plant-like life of that earth, which was still united with the sun.

So we come back to the old times of the living earth, to the times when the earth began to wrest itself from the mineral-plant life and transition to the plant. We are entering the realm of the "green rocks," which is also the home of the Labradorite. We will again meet this world of green stones in later chapters when discussing the sacred Chinese "Yu" stone, chrysolite, and diamonds. Rudolf Steiner called this time of the earth's history when these rocks began to form the Hyperborean epoch. The Greeks also knew a Hyperborea; it was the land that lay far north of them.

The labradorite belongs to those rocks that one indeed finds almost exclusively and most beautiful in the earth's north. The most famous are the finds from St. Paul's Island and the Labrador coast, from which the gemstone takes its name. Other places of discovery are Finland, Volhynia, Poland near Petersburg, Russia, and North America in New York, Arizona, Utah, Pennsylvania, Arkansas, and North Carolina.

Another gemstone in this group of feldspars is the *sunstone*, also known as the aventurine feldspar. It exhibits metallic shining light reflections with a reddish color over a bright, mostly white or even salmon-colored groundmass. Rarer are green and blue reflections. The appearance can be such that individual platelet-shaped points of light penetrating the stone can be seen, which light up and disappear again when the stone is moved. The entire surface of the ground and polished material shimmers with metallic lights as if illuminated from within. From a physical perspective, the phenomenon is based on the fact that the feldspar's crystal structure contains very small iron flakes, which can have a diameter of up to several millimeters but are only a few ten-thousandths of a millimeter thick. At this "thickness," the iron oxide (hematite), which is otherwise completely opaque, shines through with a reddish tint without completely losing its metallic reflection.

To the unbiased observer of the sunstone, the reflection from this mineral appears to be quite the same as that of the labradorite. However, one can see how the life processes underlying these mineral formations have had quite different ways of expressing the action of light in solidifying matter. Today, the living natural colors of blossoms, feathers, and hair are caused by the smallest quantities of metal combined with organic substance and are denser than the colored, transparent gemstones; the metallic is the light and color carrier. The very thin platelets of these iron oxide flakes point in general to the plant- and leaf-like nature we know from the mica.

The actual *aventurine* also belongs here (according to the light and color phenomena). These are coarse or fine crystalline *quartz* that may be embedded in slates or gneisses like veins and show the same light and color reflections as the sunstone (aventurine *feldspar*). In this case, the groundmass of the stone is *quartz*, not feldspar. Fine embedded mineral fragments and crystals cause the light reflections of the aventurine. Red and brown tones are related to iron oxide and green to fuchsite, a chromium-containing mica. Green actinolite can also produce an aventurine-like reflection. Green, white, and reddish-white avanturine come from the Ural and Altai Mountains. *The* "Tibet stone" is a rose-red aventurine from Central Asia. The Near East and Brazil also produce avanturine.

The *sunstone* was first found at the end of the 18th century on Sattel island in the White Sea near Archangel, Russia. Later on, a golden shimmering sunstone was found in feldspar veins in gneiss near Lake Baikal. The base color of these Baikal stones is clove brown. Important places of discovery are located in southern Norway near Tvedestrand and Hidra, toward the Christiania Fjord. North America has also produced sunstones from different localities.

Sunstone and Avanturine belong to a rock world (gneisses, schists), which in their layered structure begin to separate the components of granite (quartz, mica, and feldspar). In this layered structure, we can recognize something similar to the annual rings of trees. The earth as a whole had in its life processes the tendency to become tree-like during the formation of these layers.

Figure 15: Microcline var. Amazonite, with Smokey Quartz, Pikes Peak, Colorado. (Photo by Parent Géry, Courtesy of Wikimedia Commons)[x]

The *amazonite* completely falls away from this series of shimmering glimmering feldspars. Here we see only rich verdigris, which rarely has a silky, shimmering luster. Its mighty crystals sit in the already mentioned pegmatite dikes of the granite. It is chemically the same as the moonstone, a potassium feldspar. But its crystal structure is different, and it is never transparent like moonstone and orthoclase. Its color comes from a minuscule amount of *copper*. So it is again metallic, but in a place where the chemical-thinking mineralogist does not suspect it. How the copper gets into these ancient pegmatites, far away from the massive copper deposits of the earth, will always remain a mystery for the views of modern geochemistry. The actual copper gemstones (malachite, chrysocolla, azurite, and dioptase), which we will discuss later, are always found in the vicinity of the copper deposits where the copper can be mined.

However, suppose one considers that in these pegmatite dikes, the whole abundance of floral-colored gemstones is found, which owe their colors to finely distributed metals. Then it is not surprising that copper also occasionally appears there. For all its colorfulness, this metal is not one of those that play a role in coloring gemstones. The only exceptions are the amazonite and the turquoise. But copper is a metal that plays a major role in the lower animal world. In the blood of mollusks, snails, oysters, etc., it represents in the hemocyanin the iron of the hemoglobin of the higher animals.

Considering the rose quartz, we can see how in the pegmatite's tube-like cavity formations, the influence of an animal astral life came to expression. This influence may at first be flowering-plant-like as expressed by the colorations of the gemstones, but with the appearance of copper in this life sphere, it becomes apparent that a plant-animal element has entered. It is important that the copper appears in the feldspar, not in the mica or quartz. The feldspar is the lime element in the trinity of quartz, mica, and feldspar (mica, the clay-leaf-plant element; quartz, the actual mineral). We find this relation of copper to the lime element again with the mentioned lower animals and their lime casings and the copper ore deposits of the earth, which are always found where the limestone formations of lower animal carcasses cover large areas. Rudolf Steiner pointed out that wherever harmful results may ensue on the earth from the lime process, a copper process is introduced.[7]

Amazonite of special beauty is found in the Urals. It occurs there in the pegmatites together with topaz. America today

---

[7] Rudolf Steiner, *Man as Symphony of the Creative Word*, Lecture X, November 9, 1923, CW 230

supplies most of its rough stones from deposits in Colorado, Virginia, and Pennsylvania. Madagascar and Namibia are also known to have deposits. The ancient Egyptians frequently used amazonite. For the Egyptians and in the Orient, It belonged to the holy green stones. The sources for this Egyptian amazonite are unknown, although it possibly came from inner Sudan.

## 7. Precious Greenstones: Nephrite, Jadeite, Diopside, Epidote, and Vesuvianite

In the fourth chapter, we described the *"mothers of the rock crystal,"* the agate, chalcedony, and others, which were found to have a connection to an older plant-animal condition of the earth's life. According to old-fashioned but correct terminology, a rock group appears even earlier formations called "greenstones." A characteristic of these rocks is the major role *magnesium* plays in them. Even today, it is a substance that still takes part in forming the green plant pigment *chlorophyll*. In addition, the green color of these rocks comes from a special form of *iron*, which chemists call divalent iron: a fine dispersion of divalent iron silicate colors these rocks green.

These greenstone varieties – gabbro, serpentine, diorite, diabase, and others – are in their older forms interbedded with the granites, gneisses, and schists, although they transform in more recent times to trapp and basalt. Many minerals appear in the gabbro, serpentine, and diabase that catch the eye by their special structure. Already in ancient times, these minerals were given appropriate names: there is mountain wood, mountain flax, mountain leather; there is also the already mentioned asbestos, which can be made into a spinnable thread. All these minerals consist essentially of the same substance: magnesium silicate. Soapstone (steatite, powdered talcum), the stone from which magnificent stone carvings are made in the Orient, also belongs here.

The fibrous, leafy, and woody structures of these minerals and rocks, together with the predominance of silica, magnesium, and their distinctive green color, point to the *primary formative forces of plants*. They come from an epoch in which the earth as a whole tended to develop giant tree-trunk-like deposits. The history shown in the annual rings of today's tree can also be seen in the stratification of corresponding rocks such as the schists and gneisses. There, where the actual greenstones are embedded, this formative tendency is even stronger and often seemingly so woody that the layman, who sees chrysotile or serpentine asbestos for the first time, immediately speaks of the mineral as an organic structure. However, one must always keep in mind that these phenomena do not concern plant formations in today's sense, but life processes that "*embraced*" what later became mineral to a much higher degree than, for example, what is seen in today's plants with their relatively low mineral-ash content. Under special circumstances, these old life processes gave the mineral that separated from them something like an imprint of their formative forces. Even today, some plants are so strongly interspersed with minerals that they form a kind of scaffolding. If such plants, like the horsetail, are burned and ashed, a soft, silica scaffolding remains from which the plant's shape can still be clearly recognized.

The "mineral-plants," which are the basis for forming the so-called greenstones, were huge structures that were soft and gelatinous. They "swam" between other even larger formations that emerged from the earth's enormous rock layers and covered whole countries. These old, soft formations then underwent a tremendous drying and shrinking process until they became the present-day rocks.

Desiccation, shrinkage, and hardening play a special role with the greenstones. This activity is connected with the fact that the earth in this "life stage," in which it tended to become "tree-like" as a whole, really "*lignified*" and hornified its rock formations. Thus, in today's plant world, we also have the toughest, hardest, and most dried-out organic material in the woods of the trees. In the case of the greenstones, in the comprehensive sense, this is now revealed through the fact that they are not only the toughest and hardest rocks we know, but that they also contain all the minerals and metals that are considered to be the hardest, toughest, and most fire-resistant.

It is very informative when characterizing this rock group to imagine the extreme densification and hardening expressed by diamond, nephrite, jadeite, platinum, chromium, nickel, and asbestos. All these minerals and metals have their "home" in the various greenstones. Diamond, as pure crystallized carbon, is the hardest substance known. Nephrite and jadeite, which we will describe in more detail below, have a much higher compressive strength and modulus of elasticity than the best steel. The platinum group of metals has the highest specific gravities of any known substances and very high melting points. Chromium metal is so hard that it can scratch glass. Nickel is the metal that gives steel a special toughness. Asbestos, with all its fibrousness and usefulness for spinning, is nearly incombustible.

It becomes clear when looking at the properties of these various substances together with the rock in which they are "at home" that, in general, this was a formative period when the earth tended toward hardening and densification. The deeper causes of this tendency that is naturally connected with lignification will not be discussed here. The nephrite, which is found mainly in association with serpentines, gabbro, and diabase, is a remarkable mineral that is not found and used as a single-crystal like many other gemstones. Instead, it represents a true rock consisting of a closely matted and coalesced aggregate of tiny actinolite crystals. Actinolite is a mineral that is widespread as a single crystal and an aggregate in greenstones. It is related to the dark magnesio-hornblende and consists essentially of calcium-magnesium-silicate, to which small amounts of iron silicate and alumina may be admixed. Iron gives these minerals their green to blackish color.

The microscopically fine actinolite crystals forming the gray to emerald-green nephrite are as closely felted and intergrown in this precious rock as horn fibers in a cow's horn.

Tiger's Eye, Crocidolite
Orange River, South Africa
Quartz Cat's Eye (top, right)
from near Hof, Saale River

Opal in Volcanic Rock.
Fire Opal, (bottom right)
Mexico

Turquoise, in bedrock
Persia

Moonstone, Sri Lanka
(Transparent Potassium
Feldspar)

Labradorite, Labrador

Amazonite, (a variety of Microcline)
Madagascar

Figure 16: Artistic representations of gemstones discussed in this book.

Nephrite (left) New Zealand
Jade (right), polished
Myanmar

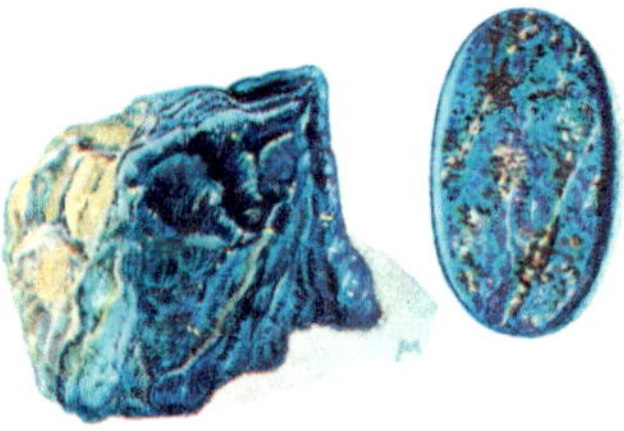

Lapis Lazuli
Afghanistan

Malachite,
Urals, Russia

Azurite
Southwest Africa

Figure 17: Artistic representations of gemstones discussed in this book.

This accounts for the extraordinary toughness of the stone. For the Stone Age human being, who did not yet know about the metals and their processing, the nephrite and the jadeite, which is still to be discussed, was a very sought-after and valuable material for manufacturing weapons and tools. The countless nephrite axes, found as traces of Stone Age cultures worldwide, are artistically worked and finely cut pieces of this precious stone. They can be seen in many museums. These stone axes, which served as weapons and tools for woodworking, are in many cases pierced and were fastened on wooden handles with animal sinew, leather, or twisted intestines.

Figure 18: New Caledonia; Battleax; Nephrite-Jade
(Photo Courtesy of Wikimedia Commons)[xi]

Large blocks of nephrite cannot be broken with a hammer. In China, where a lot of nephrite is found and processed (there it belongs to the already mentioned sacred, green Yü stones), the large pieces of nephrite are heated and thrown into cold water, causing them to crack, thereby facilitating their division.

The most beautiful, almost emerald-green nephrite comes from Siberia, from the Irkutsk Oblast on Lake Baikal's western shore. It is extracted from the alluvium of many rivers that flow into the lake. These are the best pieces, rolled by the water and freed from all cracked parts. But the gemstone is also found as outcrops in the mountains. Most nephrite is found in East Turkestan, in the high mountains of the Kun Lun range. It is quarried in stone bricks and excavated in river deposits of rubble. The stones are cloudy to translucent, of a light-green, milk-white to sulfur-yellow color.

A particularly large stone from this area is a monolithic piece from the tomb of Tamerlan in Samarkand. After China, New Zealand has the most nephrite. There, it is a translucent green color and is also known as both nephrite ore and scree. The latter is found in pieces of a few kilograms, but there are also blocks of up to 3,000 kilograms. The native Maoris have used the stone for weapons, tools, and jewelry since ancient times.

The other isolated occurrences in America and Europe are far behind in terms of mass and quality but have long been found and processed by the early cultures. For Germany, the nephrite deposit of Jordansmühl in Silesia [now Jordanów Śląski in Poland] used to be of importance.

*Jadeite* is close to nephrite in terms of its properties. However, it is mineralogically different and, above all, does not contain magnesium. The felt of crystal needles forming the jadeite consists of an augite-like mineral, essentially a sodium-alumina-silicate, to which green iron silicates are admixed. These, along with chromium silicates, give the stone its particularly beautiful green color. The jadeite is more precious and rare than the nephrite. The pieces interspersed with whiteness are much more translucent than the nephrite. There are also rosy hues caused by weathering processes, where the iron "rusts" and the original green hue changes to red.

The only jadeite deposit of real importance for gemstone production is in Upper Myanmar in the Kachin Hills. Like nephrite, jadeite is extracted directly from the mountain, where it is found associated with huge serpentine layers, conglomerates, and boulders. In the conglomerates, the boulders with other stones are fused to solid rock by a binder.

The groundmass of the jadeite is white like marble. In this white mass, emerald-green to bottle-green, often head-sized pieces are distributed, which are cut out. *Rings* and *bracelets* are made from this extraordinarily tough stone, which can be formed into delicate perforated openwork. The emerald-green color is related to its chromium content; the bottle-green color is related to iron. There are also violet hues, which are due to manganese. Rare are lavender-blue, yellow, and brown colors.

Jadeite is known to occur in East Turkestan with nephrite. In Karakash, it is said to be found together with nephrite. Jadeite is also present in the Pamir region of Tajikistan. A deposit on the Greek island of Syra may have provided the material for Egyptian jadeite scarabs.

According to findings in Mexico, jadeite also played a major role in the Maya culture of Central America. The material came from Mexico itself. In the area of old temples, thousands of jadeite beads and also jadeite tablets were found. This Central American jadeite contains some *diopside*, a green calcium-magnesium-silicate, which we have yet to discuss. The diopside makes it easy to distinguish it from the Asian-European jadeite. The prehistoric jadeite objects found in Europe are smaller and more frequently found than those of nephrite. They may have come from gemstones found in the Swiss and Piedmont Alps, but they are not nearly as colorful as those found at the Asian sites.

The just-mentioned *diopside,* with its exquisite color, is a lesser-known gemstone because of its low hardness. It is a light to dark bottle-green and also found mixed in with the oldest rocks. This gemstone is always found together with the greenstones and can appear in their fissures as magnificent crystal formations. Diopside is found together with cinnamon-brown *Hessonite* (garnet) crystals in the Mussa Alp in the Ala Valley, Italy. The color combination of long bottle-green crystals and the reddish-brown garnet-dodecahedrons provides only a small reflection of the variety of colors found in the contact zones of greenstones and various crystalline schists. Anyone who has had the opportunity to see the colorfulness in the mines of the Zillertal or the Hohe Tauern in Austria will understand the comparison with a world of flowers. A special feature of the diopside is that its crystals can be bicolored. For instance, where the crystal accumulates below, it can be deep dark green and gradually becomes colorless or almost colorless towards its tip. Such crystals are found on the Schwarzensteinalp in the Zillertal, from which the largest diopside is known.

Grass-green crystals of this mineral come from Brazil and light olive green ones from North America. In the kimberlite, the mother rock of South African diamonds that belongs to the greenstones, the chromium-rich diopside is dark green.

Figure 19: Green crystals of epidote up to 2.0 cm in length on matrix with minor manganaxinite. Found from Canta, Lima, Peru.
(Photo by var Leidus, Courtesy of Wikimedia Commons)[xii]

The *epidote* is among the lesser-known gemstones from this group of greenstones. It quickly became famous when discovered at the Knappenwand, near the Großvenediger in the Untersulzbachthal in Salzburg around the middle of the last century. This unique site yielded crystals up to 45 centimeters long and 4 centimeters thick, which can be seen today in all major museums worldwide. The color of this gemstone is pistachio green with a tinge of yellow or brown. The stone can also be completely colorless, yellow or red. In the green, yellow, and brown tones, iron is the coloring substance; the red of the so-called manganese epidote or piedmontite (after its place of discovery) comes about because manganese replaces the alumina. Epidote is a calcium-aluminum-iron silicate, so it does not contain magnesium. Therefore, it is closely related to the greenstones in its crystal formations; for example, in the Knappenwand, where it emerges from a dense mat of *byssolite,* it is known as mountain wool. This mountain wool is a magnesium-silicate, which is related to asbestos. The epidote in this locality sits in fissures of green epidote schist and is accompanied by whitish crystals of calcite, apatite, and feldspar. On some tiers of the Knappenwand, the mountain wool is matted so closely that a leather-like mineral is formed, which yields elastically to the fingernail's pressure (mountain leather).

The already mentioned *manganese epidote* (piemontite) is splendidly cherry red and is found in a manganese deposit near San Marcello in the Piedmont region. Other inconsequential localities of epidote are in Japan and Sweden, an island off Brittany, the Urals, and Madagascar.

*Vesuvianite,* a calcium-aluminum-magnesium-iron silicate, occupies a remarkable intermediate position between the greenstone minerals and the granites. The material composition of the vesuvianite, garnets, and tourmalines, which we will discuss later, is so variable and diverse that one is involuntarily reminded of the complicated composition of an organic substance. Despite this variety, you can always recognize in its crystal structure that we are dealing with a particular mineral that does not belong to any other group. The chemical complexity of these minerals alone can indicate not only the presence of inorganic activity but also those "life processes" that we have since described as the "mineral-plant world."

Vesuvianite forms beautiful crystals in predominantly green and brown tones and coarse masses, similar to jadeite. It takes its name from the earliest findings in the outcrops of Vesuvius. Another name of this stone is idocrase. The colors of vesuvianite vary greatly depending on where it is found. From the already mentioned Mussaalp in Piemont, one finds grass-green crystals; from America, yellow-brown, in Norway sky-blue (colored by copper) and rosin-colored. The egeran from the Czech Republic (western Bohemia) and the Fichtel Mountains in Bavaria appears together with soapstone; it is also a variety of vesuvianite that crystallizes radially and exhibits black to dark green tones. The jadeite-like vesuvianite (with closely fused crystal needles) comes in yellow-green to dark-green colors; there is also apple-green from serpentines in California and the gemstone pits of the Urals. The latter also comes from the Wilui River in Siberia as luminous-green wiluite with small black inclusions. In Europe, vesuvianite is known to occur in coarse masses from a limestone-serpentine contact in Bergell.

We have tried here to consider the "precious green gemstones," which should be associated with the chrysolite and the diamond because these minerals are linked in many ways to the greenstones. The home of the diamond is the kimberlite, a serpentine breccia that appeared as a transformed rock from the depths of the earth in the Cretaceous period and formed in the famous South African diamond pipes. Associated with this most precious gemstone is chrysolite or olivine, which is a magnesium silicate. In turn, this olivine is one of the most important components of meteorites, and in its variety, represents a remarkable cross-section through the realm of the "greenstones." With these facts, the events and substance formations of the earth's oldest past connect with current events, which continuously flow towards the earth from the cosmos.

## 8. Precious Ores: Hematite, Malachite, Azurite, Chrysocolla, Dioptase, and Rhodonite

What we have considered so far – the precious forms of silica, the feldspars, the remarkable greenstone minerals – have all been gemstones whose *nonprecious* forms make up the great masses of earth's ancient rocks. As was shown, the basic minerals of these ancient rocks are quartz, feldspar, mica, and hornblende. These minerals make up granite, gneiss, diorite, syenite, porphyries, greenstones, and many other rocks in various compositions and forms.

The mica group is the only basic mineral among those mentioned that does *not* have gemstone varieties. Although these silica minerals are often colored and translucent, nature could not form a gemstone from the soft leaf-like formations.

Before we move on to those special gemstones that never form rocks to a great extent, such as quartz, feldspar, and hornblende, we should recall the significant property of *metals* for endowing gemstones with color. We will encounter this fact many times. But we could also show that this property is connected with a very fine distribution of metals as they existed in the very early stages of the living earth. When the minerals and metals began to densify, minerals from the gel state and metals more from an airy state, the metals first met the abundantly present *sulfur* in the earth's atmosphere. From this meeting of the metals and sulfur arose the preliminary stages of what we know today as the metallic sulfur ores of pyrite, chalcocite, galena, cinnabar, and many others. But it would be a mistake to assume that these processes were simple chemical processes like those we work with today in the laboratory.

What existed there as airy-colored metal vapor was still alive and integrated with the living sphere of the earth. The sulfur that rose from the earth was infused with the vital warmth of earth breathing in its mineral-plant existence. Thus, the living clouds of metals encountered the living sulfur clouds of the earth.

Rudolf Steiner called what emerged from this meeting the "mother" of the later ores. But this mother substance of the metal ores was just as gelatinous as the other minerals and was now embedded in them. This embedding did not occur with arbitrary randomness but according to remarkable laws. A whole book could be written about these laws. Something of the essence of individual metals and their relationship to human beings are depicted in how metals appear in certain layers and formations. Wilhelm Pelikan revealed the secrets of seven metals in his book "The Secrets of Metals."[8] It is therefore recommended for study in this context.

The life processes from the mother substance that created the metal ores have a great relationship to processes that we can still observe today in bogs, at the bottom of the sea, and in humus formation.

For example, the so-called bog iron ore is formed in bogs. It is at first a white gel of ferrous oxide and phosphoric acid, which looks like quark. It is white and plastic, and it begins to turn blue when it comes into contact with the air. As it becomes more and more oxidized, it begins to look like rust. It solidifies at that point. If a little air is added, it becomes dry and crystalline, and the beautiful blue vivianite, a blue iron ore, is formed.

Some 500 meters below the surface in the putrid zone of the Black Sea, the sulfur of decaying animals and the iron content of the water form gelatinous *pyrite* (iron sulfide), which crystallizes very slowly.

Finally, with ordinary, undisturbed humus formation, we have processes that proceed completely in a gelatinous-colloidal state and lead to new mineral formations that *remain* colloidal.

---

[8] Willem Pelikan, *The Secrets of Metals*, Anthroposophic Press and Lindisfarne Books, 1973

(This humus formation is discussed in greater detail in the author's book "The Living Earth."[9])

The study of humus formation provides insight into the conditions of the entire living earth. One can recognize how mineral-plant and animal life (insects) were still intimately interwoven and how the present form of the living humus is only a small but necessary remnant of that comprehensive life.

Rudolf Steiner's reports from his research on the mineral-plants during earlier stages of the earth's life make it possible to understand studies of present processes. He reports that quartz, malachite, and all other minerals go back to a watery plant life already present on the Old Moon, which then repeated itself on earth.[10] After the watery plant life died out, it left behind the "mother substances" of the ores and minerals we have already described.

Here an objection is justified that many minerals and ores forming on a small scale can be observed even today according to well-known chemical processes—examples of this are given above with the bog iron ore and the pyrite.

Certainly, *these* mineral and ore formations come about through chemical processes at the present time. But the processes by which the plant builds up the material composition of its form are also chemical; only the chemistry is *different*. This plant chemistry does not work with the properties of the substances as the chemist does, but overcomes these properties and their inherent laws through life itself, to create from the simple substances water and carbonic acid sugar one time, another time wood, the third time the scent of the rose. When the plant then dies and decays or carbonizes, there is always something more or

---

[9] Walther Cloos, *The Living Earth*, Lanthorn Press, 1977

[10] Rudolf Steiner, *An Outline of Occult Science* (Also known as *An Outiline of Esoteric Science*) various publications.

less mineral, such as coal, wherein you can still see that it comes from a plant.

We can no longer see in most minerals that they come from plants. But we have pointed out those which still bear the traces of their origin. (See the previous chapter on precious greenstones.) This mineral-plant life never had a fixed solid form like our plants of today. Instead, it was flowing and constantly changing. But this mineral-plant world, which still lived completely within the fluid state, was able to do something that today's plants can also do: *it imprinted certain characteristics of its life into the substances it formed and enlivened. These characteristics express themselves today in what we call the chemical properties of substances.*

Today's plants form an immense variety of tree woods from the basic material cellulose. They create starches and sugars from the same basic material, and they also transform this basic material into resins, waxes, and fragrances. This is the way life creates substances with very different chemical and physical properties.

Just as the plant world of today builds up a variety of substances from carbonic acid, water, and light and endows these substances with many different properties *from* this life activity, so the former mineral-plant world built up a variety of later minerals from the general life substance of the earth and imprinted properties into them from *their* life activity, and these appear to us as chemistry. Rudolf Steiner characterized this general life substance as a fine fluid protein. When we observe today the new formation of minerals, we can consider these phenomena to echo once comprehensive life processes. *The chemical properties of mineral substances that we can observe today relate to the life of mineral-plants, like a dim memory of the actual event.*

After this preliminary consideration, we can now turn our attention to those few gemstones that are basically *metal ores.*

Only three metals form real gemstones in part of their ores: *iron*, *copper*, and iron-related *manganese*.

Among the metals, iron occupies a unique position because of its major role in the life of plants, higher animals, and human beings. It is well known that iron in the blood's hemoglobin is the substance that provides us with oxygen for respiration. In a similar but quite different way, iron is necessary to form the green pigment chlorophyll in plants. Thus, no other metal is so connected with the basic functions of life as iron.

This is why iron is so widely spread in the rock world and the source of life. One can say that no rock color, none of the many hues in the soils of our fields, forests, and gardens, is conceivable without iron. The life history of iron has been part of the entire history of the earth. Its so-called chemical properties have been imprinted by *life upon it* through millennia of development in the manner indicated above in the description of ancient mineral-plant life.

The mineral forms and the characteristic compounds that we can detect in the presence of iron give us a true picture of life's development since ancient times. However, it is not the place here to present these things in more detail. Whoever is interested in this special field is referred to a small work by the author in the first volume of the "Contributions to Substance Research," Natural Science Section, Goethenum, 1952: "The Mirroring of Organic Development in the Mineral Forms of Iron and Magnesium." The article shows that the formation of hematite in the mineral world can be linked with the emergence of red blood and oxygen respiration.

*Hematite* is an iron oxide, thus a compound of iron and oxygen. This mineral gives color to our sandstones, the red earth of the "Rotliegend" and the Keuper formations in western and central Europe. It occurs in this regional distribution and as a dense ore called iron glance when it forms crystals. When it appears in kidney-shaped, spherical masses, it is called kidney ore or, after an old miner's designation, bloodstone. The latter name points to the ancient use of hematite as a hemostatic agent. People once sprinkled the powdered hematite into wounds to stop bleeding.

Hematite is one of the minerals that clearly shows its origin from a pronounced gel state. Often the red ironstone (the coarse form of hematite) lines the cavities with globular kidney-shaped masses. The surfaces of these masses look like the surface of an animal organ, lung, liver or kidney, and are of dark iron-gray color with a reddish tinge. The sheen of such masses can be like a high gloss lacquer on the surface so that one wonders if it is really a natural formation. When you break this kidney-shaped ore, a fibrous crystal structure can be seen inside that usually runs radially to the surface. The fibrous structure can disappear completely if you cut a good quality piece of hematite; it can then be polished to a deep black mirror surface. The gemstone has approximately the hardness of feldspar and can scratch glass.

Cutting this stone requires a special skill to bring out all the material qualities despite its fibrous structure. The cutter must carefully look at the stone before starting work to understand the surfaces and curves. Since it is opaque, everything depends on the right workmanship of the deep black shiny surfaces. In this work, something else occurs that can at first frighten the visitor to such a hematite cutting shop: the fine powder of hematite is cherry red, and since the stone is cut wet or on a grinding wheel under running water, the grinder and the workshop look like they are stained with blood.

This rock's greatest quantity and most beautiful material come from an iron ore mine in Cleator Moor (Cumberland). Other occurrences are on the island of Elba, in Brazil, India, North America, New Zealand, Norway, Sweden, and Germany. Hematite is particularly suitable for engraving signet rings. It was already used in Egypt, Babylon, and Greece for rings, gemstones, and pearls.

Copper is the second metal that can appear in its ores as a gemstone. While *iron* has a special relationship to the oxygen, weathering, and decomposition of its ores, *copper* has a special relationship to *carbonic acid,* which can be beautifully observed in everyday life when looking at copper-clad roofs. The original reddish copper is quickly covered with a green *patina* of copper carbonate (verdigris). While iron rusts, copper covers itself with a patina that prevents further exposure to the air.

The most important of the more precious copper ores is *malachite*. This gemstone is made of copper carbonate containing about eight percent water; it has a dark emerald-green to black-green color. Like the hematite, it appears in kidney-shaped globular masses, whose outer roundish surfaces are mostly black and dull. Thus, at first, one does not see the real color of the stone.

Figure 20: Malachite and azurite from Morenci in Arizona, United States. (Photo by Marie-Lan Taÿ Pamart, Courtesy of Wikimedia Commons)[xiii]

The emergence from a gel state can also be seen in the globular forms, which may have developed into grape-shaped and stalactite forms. The stone always has a radial fibrous structure in its interior, but this is joined by a concentric shell-like structure parallel to the surface, similar to agate. This “woody texture” of the malachite perpendicular and parallel to its surfaces often gives intersected stones the appearance of wood. The individual fine layers lie in changing shades next to each other like a tree's annual rings. This arrangement clearly shows that the whole mass gradually forms, layer by layer, as a soft, plastic substance, then crystallizes radially into fibers during solidification. The formation took place in subtle rhythms, gently forming plant-like from within. Beautiful, diverse effects come to light through the cut of this relatively soft stone. The so-called “eye” malachites come about by a rounded grinding of the globular formations that exhibit lighter concentric layers around a darker core. Additionally, malachite with a silky luster, the so-called “shimmer” malachite, comes from California.

Closely related to malachite is copper lasur or *azurite*. Although it is also a copper carbonate, azurite occurs nearly always in crystalline form. These crystals are deep cobalt-blue and transparent with a strong surface luster. The azurite has somewhat less water in its crystals, about five percent, and can change into malachite through water absorption. Thus, in Bisbee, Arizona’s copper mines and the famous Otavi mines in Southwest Africa, azurite is found with crystals completely or partially transformed into malachite. These are so-called pseudomorphs of malachite to azurite. The transformation process is also associated with a small loss of carbonic acid. The azurite crystals become green, completely opaque, and then consist internally entirely of fibrous malachite through this process.

Azurmalachite is a random or layered mixture of azurite and malachite, which occurs in grape- and kidney-like masses. Spherically cut pieces of such material expose intensive green and blue colors similar to the eye of the peacock’s feather.

The most famous sites for malachite are located in the Urals in the area of Yekaterinburg. Kidney-shaped and stalactitical masses up to 1500 kilograms were found in the mines of Gmeshevskoe. In 1855, a mass of 25,000 kilograms was found in a mine near Nizhny Tagil. These malachite masses were used in Russia under the tsarist rule to make artistic crafts, chandeliers, bowls, vases, fireplace mantels, tabletops, etc.

In St. Petersburg, the columns of St. Isaac's Church were entirely covered with malachite. In this context, it is worth mentioning that the malachite used to cover the Hagia Sophia's columns in Istanbul was formerly at the Temple of Diana in Ephesus, Greece. Where this malachite came from is unknown. Significant quantities of good quality malachite for jewelry and craft items come today from the large copper mines in the Katanga Province, the Democratic Republic of the Congo. Copper mines in Australia and Chile also produce malachite.

Azurite often occurs together with the malachite but is found mainly at Chessy near Lyon in France.

The *chrysocolla* is a copper silicate gel found in green or blue-green, grape-like masses covering copper ores or as a fissure filling in copper ore deposits. Despite the silica content, the stone is very soft since it is still completely in the hydrous gel state. It is often found together with malachite in the same deposits. Noteworthy is a site at Lake Superior in North America, where the chrysocolla is mixed with quartz and thus somewhat harder.

The *dioptase* is also a copper silicate but fully crystalline and transparent. The name copper-emerald characterizes the beautiful color of this gemstone. It has about the hardness of glass. Although it is a silica compound, it is remarkable that this mineral is found almost exclusively in limestones and dolomites. (The dolomites are calcium-magnesium carbonates.) Observing specimens from many localities makes it possible to see that the green crystals of dioptase are formed suspended in a gel of limestone or dolomite. Only later did the mother substance change into a crystalline state.

The most famous locality for dioptase is at the Altyn-Tyube mountain, Bukhar-Zhyrau, Karaganda Region, Kazakhstan. Pieces worth polishing now come from the copper mines of the Guchab-Rodgerberg area in Namibia, Clifton in Arizona, Băiţa mining district (Rézbánya) in Romania, Copiapó in the Atacama, Chile, Cordoba, Argentina, and Comba, Ngamaba (Brazzaville).

*Rhodonite* is a manganese silicate of a dark rose-red to bluish-red color. Crystals of this mineral are rare. For jewelry and craft purposes, a dense, fine crystalline rhodonite often found interbedded among clay shales is used. Manganese is closely related to iron and has played a major role in the mineral-plant world of the earth's past. The occurrences of manganese ores are particularly associated with the shales and thus show this metal's relationship to the "tree stage" of earth formation, which we have mentioned in the chapter about the precious greenstones. Even in today's plant world, manganese is still very widespread and can be easily recognized in the ashes of many plants because it gives them a green color when they are thoroughly fired. The coloration of manganese is very striking in its salts. There are the deepest violet (potassium permanganate), green, and rose-red. In the mineral world, most manganese ores are deepest black, only in the manganese carbonates, and in the rhodonite, does the rose-red appear. We recall here the rose quartz, which also owes its color to manganese.

The massive rhodonite deposits make it possible to saw the rock, which can be used as a splendid cladding material. The rose-red is often cloudy and streaked with black veins of manganese oxides. Famous localities are near Yekaterinburg in the Ural Mountains, where a layer about three meters thick is mined in a black clay shale, and at Langban in Sweden at the manganese ore mines there. From Massachusetts in North America comes a pure flesh-colored rhodonite that is very popular in America. Other sites are in California and Australia.

## 9. Lapis Lazuli and Turquoise

With lapis lazuli and turquoise, we not only touch upon certain secrets of rock formation in the world but above all on the most ancient, advanced prehistoric cultures of humankind. It is remarkable that these two gemstones, which are not related materially, are found almost exclusively in areas of ancient cultural centers of Central Asia, the Nile Delta, Central and South America. According to findings, both stones have played a prominent role in ancient Egypt's ritual life as early as 4000 BC. Mexican turquoise has been found in the remains of the Inca, Aztec and Maya cultures of South and Central America. The extent to which the Chilean lapis lazuli, found at 3500 meters above sea level in the Andes mountains, was known to the Incas is still undetermined.

Because many gemstones are found in clusters in certain areas of the earth, it is necessary to consider the relationship between the formation of the earth's surface, the characteristics of its rocks, and the cultures that developed upon them.

It is commonly believed that the formation of the earth's surface is purely accidental and has no connection with the later development of certain cultures. However, Rudolf Steiner presented ideas about the mutual development of the earth and humanity in his cosmology which gives us the possibility, together with scientific and archaeological research, to form ideas about the connection between what is below the earth's surface and the great course of cultures. Continuing Rudolf Steiner's suggestions, Gunther Wachsmuth made essential contributions to this subject in his books "The Evolution of the Earth" and "The History of Mankind." These works are explicitly referred to here.

In his book "An Outline of Occult Science," Rudolf Steiner writes about an ancient culture or race that had already reached a certain degree of physical-material existence towards the end of its time, only to perish in catastrophes of fire. This Lemurian culture took place exclusively in the southern hemisphere. Rudolf Steiner defines the "continent" of this culture as "situated" between South America, Australia, South Africa, and the Indian archipelago. It is that "continent" which modern paleogeography calls Gondwanaland. The actual existence of this "continent" for a relatively short time in the history of the earth is now confirmed by extensive research in geological, botanical, and zoological fields. This research has shown that the formation of the seabed where it was located, the adjacent coasts, and the plants and animals in South America, Australia, South Africa, and India point to such an ancient continent.

Before the demise of Lemuria, a new continental area began to develop between South America and Africa, which gradually unfolded into the northern Atlantic Ocean. Since ancient times, this place has been known as the Atlantean continent, about which Rudolf Steiner spoke extensively in his books and lectures. The southern, warm part of this continent appears in the Germanic Edda as "Muspelheim" (the realm of fire), the northern, cold part as the well-known "Niflheim" (the world of mist). However, before we go into the cultures that originated from this Atlantean continent, we must once more turn our gaze back to the declining Lemuria. The "fire catastrophes," under which the disappearance of this continent took place, are well known to modern geology as that strong "volcanic" activity present towards the end of the so-called Triassic, during the Jurassic, and above all in the Cretaceous period into the Tertiary over large areas of the earth. The peripheral areas that were described above as the approximate contour of the Lemurian "continent" are at the same time *the gemstone sites* of the earth: India and the Indian archipelago, Sri Lanka, Madagascar, South Africa, Australia, and South America. The gemstones found outside this ring pale compared to the brilliance and beauty of the sites in the indicated areas. The Urals is the only exception.

How is it to be understood that just in these marginal areas of the old Lemurian "continent," which later become carriers of large, early cultures, gemstone finds are prolific?

That evolution of the living earth from a more plant-animal to a plant state, which we have mentioned repeatedly in the preceding chapters, took place at the beginning of the Lemurian period. Towards the end of this epoch, a large part of the substance appearing later as rocks and minerals has already been eliminated from the unfolding life processes of the developing nature realms. *However, it has not yet solidified* and is still in that gel state that has been frequently described here. Spatially the earth was still many times larger than today because its atmosphere was filled with gel and extended up to a height of 300-400 kilometers into higher airy-watery layers. Inside this gel, there were already more solid conditions, densifications up to the hardness of horn, and around which flowed fluid matter. Out of such currents, the continents were formed, breaking up and densifying again.

Rudolf Steiner describes the human being of that time as not yet visible in a physical-material body. However, toward the end of the Lemurian epoch, human beings were already walking upright. They begin to form their physical bodies, especially regarding their sense organs. This process is only completed in the middle of the Atlantean epoch, the cultural epoch that follows the Lemurian.

Suppose we characterize the spiritual situation towards the end of ancient Lemuria with a picture. In that case, one might well imagine that all the spiritual powers were "gathered" around this "continent" and had the task of establishing the human being as a spiritual, ensouled, and physical creature on the earth as it is becoming mineral.

Before this point, humans were beings swimming like an embryo in the amniotic fluid of the earth; now, the "birth" began by detaching from the great "placenta" of the earth. Thus, human beings begin to clothe themselves with skin, separate themselves from the environment, and enter into connection with it again in a different way through the developing sense organs.

As we already reported in chapter "2. The Rock Crystal," the formation of the skin and the sense organs are intimately connected with the silica process. Rudolf Steiner describes in a lecture[11] that the formation of the sense organs in human beings is *paralleled* by a corresponding process in the mineral realm, which leads to the formation of gemstones. In this sense, those relations between gem formation and sense abilities should be understood. The formation of the sense organs and their capacities *begins* toward the end of Lemuria. *Simultaneously, in and around Lemuria*, a diversity of gemstones begins to develop in the emerging mineral realm we have characterized. The human being is here the "measure of all things."

It may be mentioned here that Rudolf Steiner described the island of Madagascar as a "remnant" of Lemuria. This island is an earth region that did not experience such large transformations of its life and rock existence as did other regions. In this context, it is significant that in Madagascar, in contrast to nearby South Africa, a larger variety of particularly precious gemstones are found.

After this excursion into the deeper connections between human beings and gemstones, we can now follow the traces that lead us to the sites of lapis lazuli and turquoise.

It was pointed out above that the Atlantean continent was located between America and Europe-Africa. It was the carrier of many great cultures, which Rudolf Steiner described in his book "Cosmic Memory." Before this continent's demise, smaller groups of certain races, led by outstanding personalities, emigrated from this region to Central Asia and Central America. Central Asia became the cradle and starting point of the great Asian-European cultures: India, Persia, Babylonia-Chaldea-Egypt, Greece, Rome, to

---

[11] Rudolf Steiner, *The Relationship of the Human Sense Organs to the Outside World*, Berlin, October 19, 1906.

the present. Central America became the starting point of the Mayan, Aztec, and Inca cultures.

As already mentioned at the beginning of this chapter, Central Asia, the Nile Delta, Central and South America are also areas where lapis lazuli and turquoise are found. One must first allow such a fact to exist. The ritual use of these minerals only partially clarifies the riddle of the connection between certain cultures and special mineral formations in their region. A solution to this riddle can only be sought by realizing the special nature of these gemstones, of which only the priests of these cultures had an insight. The use of these stones as ritual ornaments and utensils expresses pictorially something of the inner natures and tasks of the cultures in question. In its ritual use, the stone was *not a symbol* when used in cultic rituals but a direct expression of the divine working into the darkness of the solid mineral matter.

To better illustrate this, let us first consider *lapis lazuli*. The most remarkable thing about this opaque deep ultramarine blue gemstone is that its color does *not* come from a dispersed metal, as is the case with almost all other gemstones. From a chemical perspective, it consists of three different minerals, which occur in variable mixtures. Two of these minerals, the haüyne, and the sodalite, occur in many other rocks, but they do not form lapis lazuli. For the lapis lazuli, a third substance is needed, which is nothing other than the well-known lazurite (ultramarine) produced artificially after chemical studies of the blue, lazurite-like formations in glass furnaces.

What is the actual material of lapis lazuli? One of the minerals is sodalite, a sodium aluminosilicate that contains colloidally distributed table salt. The haüyne is likewise a sodium aluminosilicate, but it contains finely distributed gypsum. Both minerals can be blue *by themselves* because of the finely distributed salts. The third substance is the actual lazurite, which is chemically also a sodium aluminosilicate. But, in contrast to the other two minerals, it contains a colloidally distributed sodium sulfide. Sodium sulfide is itself a substance that rapidly dissolves and decomposes in the air or is oxidized. However, its fine distribution in the rock has largely escaped the usual chemical laws and exhibits a blue color, while sodium sulfide is a yellowish-white salt.

Thus, we have an extremely complicated substance mixture before us whose individual constituent parts are well known. We can even manufacture one component – lazurite – by artificial means. But this very fabrication, which takes place at more than 1000 degrees Celcius, tells us that in nature, lapis lazuli could *not have been formed at such high temperatures* because it was formed in marble-like limestones and is also permeated with this limestone substance. It often contains many bright golden-yellow shining sulfur granules, which in large polished stones look like stars in the blue sky. These Lapis Lazuli containing limestones are accompanied or surrounded by granites. Therefore the gemstone is referred to as a formation occurring in the contact zone between limestone and granite.

Lapis lazuli very rarely forms crystals; it has a granular structure like its parent rock, limestone. These accompanying circumstances let the lapis lazuli appear as a formation developed from the very complicated relationships of force and substance. There is no doubt that it was formed relatively late and that its substance was for a long time in that colloidal, warmed state of suspension characteristic for younger minerals. Due to these special relations of substance and force, it was possible that a substance could come about that turned one of the most earthly substances, namely sulfur, into its opposite. Normally sulfur is yellow, but it appears here in its opposite color, purest blue, as a fine distribution in a silica mineral related to the feldspars. There should be no other substance in the mineral realm about which one can say that its original earthly properties are so transformed by the influence of opposite forces that something of the region from which these opposite forces originate becomes visible: in this case, the pure blue of the sky surrounding the earth. In the laboratory, quite complicated processes must be carried out to transform the yellow sulfur into its blue form. Sulfur in Latin means “sun bearer.” In lapis lazuli, this “sun-bearer” is not present in the form we know, for example, from Sicily’s sulfur pits or from the emissions of the great volcanoes: that poisonous yellow sulfur whose vapor inhibits life. The “sun-bearer” is present in this gemstone in its original form, as blue, cool light, as it once surrounded the earth in clouds in the distant past. The earth’s forces have made this light substance into a very important but also very dense material. Only in the living protein of plants, animals and humans does sulfur still exist as it did in the earth’s ancient past. Lapis lazuli thus embodies something of how the sun's forces can connect with the earth. The “sulfur” contained in it is not the normal earth sulfur but is still the original “sun carrier.” The special thing about this phenomenon is that such an unstable and active substance as sulfur appears in nature in an extraordinary variant. We can fabricate such variants from many laboratory substances, but they do not exist in nature. In this respect, lapis lazuli is unique.

Connecting the sun's forces with the earth existed as a powerful hope for the future in the religious life of Persian and ancient Egyptian cultures. It was these two cultures that worshiped and used lapis lazuli. However, the ancient Indian culture, which was still quite remote from the world, did not know sun worship in this sense, and lapis lazuli is not to be found in the Indian rituals.

Nevertheless, the lapis lazuli is found in the Indian cultural region, in Myanmar and Afghanistan’s northwest corner. Another important occurrence is at the western end of Lake Baikal, north of the Gobi Desert, which may be considered the center of that Central-Asian cultural cradle of which we have spoken above.

The Egyptians made their sacred scarabs mainly from the lapis lazuli. The scarab is the artificial replica of a beetle that inhabits the Mediterranean countries as the "Scarabaeus sacer." It was a symbol of the sun and heart among the Egyptians.

The *turquoise* is materially and, regarding its appearance, something completely different than the lapis lazuli. The pure sky-blue to greenish-blue stone is also opaque and is found in Mexico, southwestern North America, and Persia in the fissures of strongly decomposed volcanic rocks. Other occurrences with unknown geological circumstances are found in Tibet, Afghanistan, and near Samarkand. The ancient mines in Egypt, mostly on the Sinai Peninsula in the Serbal Mountains, are located in sandstone from the Carboniferous period and are covered by basalts. The Egyptians exploited these latter deposits as early as the *fourth millennium B.C.* They have not yet been exhausted.

The turquoise is materially a very complex hydrated copper-aluminum-phosphate, which can be delicately impregnated with silica by the accompanying chalcedony. The relatively soft stone sits as a globular (botryoidal) cover in very thin layers, often only a few millimeters thick, in the cracks of weathered rock. Its inner structure is granular or fibrous and consists of microscopic crystals—the spherically-globular formation of the turquoise crust on the rock points here to an original gel state. Depending on the amount of this original gel still present in the turquoise, the stone's water content and color consistency vary.

Recent research has shown that all turquoise contains ammonia, a substance that we know very well. It occurs in nature only by decomposition of organic substances and the emissions from hot springs and volcanoes. The ammonia has a special relationship with copper and forms splendid intensely blue-colored compounds with the salts of this metal.

A remarkable combination of substances and forces in the turquoise would have been possible in many places on the earth, but only in a few did it lead to turquoise gemstone formation. The basic substance of turquoise is an aluminum phosphate that chemically and materially could form everywhere rocks containing aluminum-phosphate weather and decompose. Therefore, aluminum phosphates are relatively widespread, especially where minerals containing sulfur were "burned" by volcanic warmth and where the sulfuric acid formed attacked clay rocks. Yet, this chemical distribution does not "explain" the material composition of the turquoise. An attempt to artificially create turquoise based on these ideas about its formation doesn't work. On the other hand, aluminum phosphates are still formed today through the weathering processes mentioned above, but these are not turquoise, and these phosphates do not have the structure of the old forms.

These facts always lead us back to the earth's past, where the properties of these substances must have been completely different, as we have shown several times in previous observations.

We have already mentioned the clear gel structure of the turquoise. It is the same as in malachite, hematite, and others: an original gel has segued into the crystalline state. It is possible to have beautiful ideas about the formation of the turquoise, but it is not possible to make turquoise without restoring this gel state. The gel state of many substances can be produced today, but it cannot be preserved for a long time. Most substances *today* tend to crystallize very quickly from the gel state. This condition was not the case in the earlier life states of the earth because the substance was more animated then. We have already mentioned earlier that the gel state is *the* state of living matter.

The turquoise has always formed in the vicinity of "volcanic" processes. But there is also a metal found almost exclusively in connection with such "volcanic" processes. This metal is *copper*, which gives the turquoise its unique sky-blue color.

Let's look at everything necessary to make the turquoise formation possible. First, we can see that the metal "coloring" the turquoise was already in a much denser state than we know it from the transparent gemstones. Copper is lifted out of its earthly densification through the ammonia-nitrogen process, which is connected with the airy regions of the earth. The phosphorus strengthens this process. The copper is connected with the aerated and light-filled phosphorus and then with that substance from the old mineral-plant life of the earth: alumina. Suppose one also considers that this entire "process" took place in such regions of the earth where something of the old life was still present in the active "volcanism." In that case, it is possible to recognize the turquoise as a structure that emerged from the earth's last dwindling life processes. As with lapis lazuli, something extraterrestrial is interwoven with the colored copper of the earthly; the sky is once again depicted in the dying earth. In this sense, the two gemstones, lapis lazuli, and turquoise are related. Once again, heaven is connected with the earth, but it no longer achieves transparency like clear, crystallized gemstones.

The character of the two gemstones, which we tried to explain here using present-day knowledge, may have been the "reason" both were held sacred by those past cultures that hoped to reunify the spiritual and the earthly in the future.

## 10. The Topaz

With our previous considerations, we became acquainted with the ancient life conditions of the earth, which could be called "flower-like" after the indications of Rudolf Steiner. We found mica and feldspar as the remnants of "inwardly proliferating" flower processes forming the earth's life at that time. Together with the quartz, these minerals form granite and similar rocks, so we can say that this granite is the transformed exudate and remnant of ancient life. More recently, younger states of earth life have exuded shale, coal, and limestone, similar to peat formation in the moors. But the ancient life condition of the earth was neither animal nor plant alone. The mineral, plant, and animal interpenetrated in a relatively primitive life process and therefore had this "inwardly proliferating" character.

During our examination of the rose quartz, we met with a phenomenon in the granite called *pegmatite*. As we described, these are cavities in the evenly grained rock in the vicinity of which an extraordinary increase of individual minerals in the rock can occur. Inside these cavities, the most beautiful minerals are formed as crystals. We find rock crystal, smoky quartz, feldspar, and hexagonal, closely packed mica crystals that can be split into many single, transparent lamellae. But an abundance of more precious and non-precious minerals, which are otherwise finely distributed and imperceptibly hidden in the granite mass, suddenly appear in these cavities as substantial crystals. From this abundance, which cannot be treated exhaustively within the framework of a small study of gemstones, we want to select only the *topaz* at this time. Later we will see that other gemstones are also at home in these hollow spaces in the granite.

The crystal community in the pegmatites is a very interesting phenomenon. It shows that the most diverse substances or chemical compounds of silica, aluminum, potash (potassium carbonate), fluorine, and phosphoric acid have *developed* side by side peacefully without disturbing each other's "development" through their material differences. One can compare this fact with another one, that in a meadow, innumerable grasses, flowers, and herbs of the most varied kinds grow together without disturbing each other. The coursing life of individual plants, at first invisible, incorporates densified substances from air, water, and light into their hidden gestures and thereby become visible. The growth and form of the plants result from substance-filled movements.

Crystals develop like plants. As we have already seen with the rock crystal, they can even have different "habits" in different regions of the earth. Their shape corresponds to invisible gestures to which the earth gives a relative duration when filled with matter. The formative forces of the various crystals are still present today, but that "living" state of a substance that filled them is past. "The formation of crystals is the reaction of the earth to the cosmos," said Rudolf Steiner in a note from 1923. In a later brief consideration of the so-called synthetic gemstones, we will see that human beings today can technically produce "imitations" of such "living" states and actually crystallize gemstones. The juxtaposition and the succession of different crystal formations in the druses of the pegmatite show us that the individual principle we have already spoken about prevails just as it does in the plant world. If carbonic acid, water, and light produce oak wood one-time and another time birch wood, so from silica, alumina, and potassium carbonate come one-time feldspar and another muscovite.

Just as the plant today still condenses most of its material composition from the air, water, and light, thus from a transparent, almost invisible state, so the life of the earth exuded *all the substances* we have today in the mineral realm from similar airy, luminous, and warmth states in the most ancient past. At that time, it was first gel-like and only later became crystalline. The human being's development was also intimately connected with this life of the earth. We have already pointed out at the beginning that the formation of gemstones was closely connected with the formation of sensory abilities and sensory organs in human beings. This idea is a connection that cannot be "proven" only through scientific observation. Rather, it is a hidden connection that only comes to light when a person gets to the essence behind the world of appearances, as Rudolf Steiner has done. In this region, the human being encounters as their first experience the invisible warp and weft of the world's visible carpet. Only where the threads are interlaced and interwoven and cross each other according to hidden laws does something appear visible to the earthly eye.

By themselves, these connections between gemstones and sense organs have been known since ancient times. Rudolf Steiner pointed out that they can still be found correctly in the tradition of the medieval scholar Agrippa von Nettesheim, while they appear false and confused in many other occult writings. He has investigated them anew on his modern path of knowledge and spoke about it in October 1906.[12]

There it is said that the *topaz* is connected with the *sense of taste*. This sense is centered in the more or less conscious part of the tongue. Its unconscious part reaches down to the metabolism, where the liver is the organ dealing with the *inner* taste of food that comes to consciousness again with the *digestibility* of the food substances. So the tongue and the liver are actually the two poles of the comprehensive sense of taste.

In the vicinity of the tongue are the teeth. They have little to do with taste; at most, they enhance the taste through the chewing process. However, *one* substance plays a tremendously important role *in* the teeth and has become better known in recent years: *fluorine*. This fluorine is present as an apatite-like substance (as fluorapatite), mainly in the tooth enamel. It also works from there into the metabolism, where, according to Rudolf Steiner's research, it leads to the purposeful and agile movements

---

[12] Rudolf Steiner, *Esoteric Development*, Lecture IX: Imaginative Knowledge and Artistic Imagination, October 21, 1906.

of the limbs. In this entire fluorine process in the human being, the teeth form something like suction organs through which the fluorine enters the organism. Another very important substance in this process between the tongue and liver is *tin*. In medical lectures,[13] Rudolf Steiner presented the metals and their relationship to the different organs. The liver is referred to there as the tin organ. This statement is not to be understood as meaning the liver contains tin but that the underlying tin *dynamic* is active in the liver. This dynamic of tin lies in a remarkable relationship with solid and liquid states. For those who are more interested in the inner forces of the metals (which play an important role in anthroposophical medicine), the book by Wilhelm Pelikan, "The Secrets of Metals" and Gisbert Husemann's *Erdengebärde und Menschengestalt* ("Earth Form and Human Form," untranslated) are recommended for study.

Now, however, in the course of more detailed studies of the human body concerning metals, it has been found that the *tongue* contains a particularly large amount of tin, indeed the most tin of all the organs in the body. This fact is one of the innumerable established facts in science that now lie like a brick in a great heap of bricks, without the possibility of fitting these bricks into the world's great structure in a meaningful way.

In connection with the sense of taste and the tin content of the tongue, we want to try for once to fit in such a lost brick. We have already shown earlier that silica plays an important role in the *entire* realm of the senses. In the realm of taste, silica is joined by the fluorine-calcium process of the teeth and the mysterious tin in the tongue.

These substances that appear in the human being – silica, fluorine, lime, tin, and the phosphoric acid in the bone-like calcium fluoride phosphate of the teeth – form a *characteristic community* in the granites of certain regions outside in *mineral*

---

[13] Rudolf Steiner, Spiritual Science and Medicine, March 21st to April 9th, 1920, Dornach.

*nature*. This community is well known to mineralogists and contains cassiterite (tin ore), topaz rock, fluorapatite, and fluorite (calcium fluoride). These are all the substances playing a role in the region of the human sense of taste, and *topaz*, which is in the middle of it, is a *fluoro-aluminosilicate* and chemically composed of silica, aluminum, and fluorine. The alumina is the only substance that does not appear in the human sense of taste. Nevertheless, it plays a mediating role, as everywhere in nature, between silica and calcium.

Such an example can show how the order in nature has its cause and measure in the *human being*. Future natural science will have to increasingly look for this human cause and measure in nature because everything that happened to and in human beings during their long process of development has its counterpart outside in nature.

Figure 21: Topaz cluster, Topaz Mountain, Thomas Range, Juab County, Utah, USA. (Photo by Rob Lavinsky, Courtesy of Wikimedia Commons)[xiv]

Now the topaz itself, which sits there with its companion minerals in the hollows of the pegmatites, is one of our hardest gemstones. It has a hardness of 8 according to the Mohs scale; thus, it is harder than the rock crystal with 7. It has a very wide color range. In addition to a predominantly yellow color, which plays between the most delicate light to dark brownish yellow and often with a reddish tinge, there are many soft blue stones, sometimes with a greenish tinge, and many completely colorless stones that are called "water drops." Rare is a red that plays between light rose red and delicate lilac.

Metals determine the color of topaz: chromium gives the yellow tones, iron the blue tones, and the rare iron-related vanadium the red tones. Vanadium is a metal that plays an iron-like role in the blood of lower animals and is also found in abundance in the ashes of certain coals. It may therefore have had greater importance in earlier states of life on earth. Vanadium is also found in human bone.

Figure 22: Colorless topaz, Shigar Valley, Pakistan.
(Photo by Géry Parent, Courtesy of Wikimedia Commons)[xv]

Topaz crystals are distinctive and appear similar to the rock crystal and exhibit a typical "habit" that characterizes its locality. It is necessary to be very careful when loosening topaz crystals because the topaz has a particular cleavability parallel to its base. Each crystal detached from its connection point has a flat surface at the bottom, allowing it to be easily positioned. The size of the crystals can reach 25 cm long and 10 cm thick. Good, polishable stones are usually smaller.

The sites of this gemstone are not extremely widespread. However, very beautiful blue stones come from Mursinka, Russia, colorless ones from Miask on Lake Ilmen, Russia, and yellow, pink, violet-blue ones from the gold placers of the Sanarka River in the Southern Urals. In Transbaikal, especially on the Urulga River, beautiful and large topaz of various colors are found. The crystals range from 3 to 30 pounds in weight. In addition, green topaz comes from Kamchatka.

Figure 23: Blue Topaz, Itinga, Jequitinhonha Valley, Minas Gerais, Brazil. (Photo by Rob Lavinsky, Courtesy of Wikimedia Commons)[xvi]

Brazilian topaz deposits are also significant. They are located in Minas Geraes and yield colorless, yellow, reddish, and blue crystals. In some places, the crystals are also found in weathered rock stripped from old rock formations. There the crystals appear intact, embedded in clay and marl.

In gemstone placers, such as those in Sri Lanka, topaz can be found completely rounded by water.  Numerous other sites are located in Asia, Africa, Australia, North America, and Europe, without reaching the importance of the Russian and Brazilian sites. Worth mentioning is the topaz in the tin region of Schneckenstein, Saxony, southeastern Germany. The Saxon kings exploited beautiful yellow and yellow-green polished stones now in the Green Vault Museum in Dresden.

## 11. The Family of Beryls: Emerald, Aquamarine, Chrysoberyl, Euclase, and Phenakite

The occurrence of topaz is in a narrow sense closely related to that of the *beryls*. But, as we shall see, this large family of gemstones yields enormous crystals from its non-precious varieties and exhibits a very characteristic structure, the meaning of which only becomes clear when the relationship of these gemstones to the human being is considered.

It is important to distinguish between the simple beryls – to which the aquamarine belongs with the related phenakite, euclase, and emerald – and the chrysoberyl with its variety alexandrite. Beryllium, from which the beryl derives its name, is a rare lightweight metal that all of these gemstones have in common. This metal has a great similarity to aluminum; only it is even lighter. It is scarcely twice the weight of water.

We have seen that alumina (from which aluminum is made) was originally found mainly in mica and feldspar and played a major role in the mineral-plant world of the earth's past. The beryl ore is much rarer but is likewise completely connected with this old mineral-plant world. The only difference is that it is exclusively linked to the essence of certain gemstones.

The material of the beryls is a beryllium-aluminum-cyclosilicate, in other words, a silica compound. Different finely distributed metals bring about the numerous varieties of this gemstone.

The *emerald*, with its unique color of purest green, is tinted with chrome. The *aquamarine* owes its seawater-blue to sapphire-blue coloration to iron. The *rose beryl* or *morganite* contains the rare alkaline element cesium, which does not color it; the metals manganese, nickel, copper, and iron together give it its beautiful peach-blossom salmon color. The common light green to yellow *beryl* contains iron in a different form. Then, the so-called golden beryl or heliodor is greenish to bright golden yellow and colored by uranium. This and zircon are the only gemstones in which uranium has been found.

*Euclase*, found along with some beryls, is a hydrous beryllium-aluminum-silicate, colorless, or colored blue-green by chromium and vanadium. It is very rare.

*Phenakite* rarely accompanies beryls and emeralds, but It is only a beryllium silicate and does not contain aluminum. Mostly it is colorless; brown, yellow, and red tones are rare.

Figure 24: Chrysoberyl with a star form, Minas Gerais, Brazil
(Photo by Matteo Chinellato, Courtesy of Wikimedia Commons)[xvii]

*Chrysoberyl* and its close relative *alexandrite* are two very remarkable gemstones that are a compound of alumina with beryl ore containing *no silica*. Nevertheless, these two gemstones are harder than the usual beryls mentioned above (7) and harder than the topaz (8); they have a hardness of 8.5. Chrysoberyl commonly has a very high iron content that, together with other metals, causes the stone to have an extremely wide range of colors. They range from light yellow through several gradations of green to dark green and greenish-brown. Two remarkable features can be observed: the *chrysoberyl cat's eye* or *cymophane* exhibit a silky, more or less contoured shimmer of light, which is caused by numerous fine channels inside the crystal (up to 25,000 per square centimeter) and the alexandrite, which is always a more or less dark to emerald green chrysoberyl (by day in sunlight), but exhibits blood- to violet-red color under artificial illumination (especially candlelight), which can increase to ruby-red in good stones.

Rudolf Steiner has indicated that the gemstone family of beryls is closely related to the lesser-known senses of the human being. He connected *beryl* with the *sense of thinking* and the *emerald* with the *solar plexus* (unconscious movements in life). [xviii] This differentiation of the beryls in relationship to the human being throws light on the quite different nature of beryl and emerald. As we indicated above, one must distinguish between the actual beryls: aquamarine, rose beryl (morganite), common beryl, heliodor (golden beryl), and emerald. *Except for the emerald, these gemstones are found only in the cavities of granitic pegmatite, just like the topaz.* Because this group of gemstones has a special relationship in its history to the more conscious part of the human senses, it is understandable that the beryl (sense of thinking!) is found there.

The situation is different with the *emerald*; even if it is beryl, it is found exclusively in *schists*, mostly in mica-schists and younger shales. These schistous rocks are in their origin much more *plant-like* than the uniformly granular granites, which arose from a still quite undifferentiated, simple, floral life of the plant-animal. We have pointed this out in the case of the more precious feldspars. These schists have a special relationship to the "plant-like" processes in the human being, mostly in the middle and lower human being (solar plexus, autonomic nervous system).

However, the emerald has another special feature that I would like to point out. Unlike most other gemstones, the emerald has not yet reached its full hardness. When freshly found in the interior of mica schist, it has peculiar strings of crystals running irregularly through the schist (emerald does not form crystals in cavities like the other beryls). It is so soft that a person can sometimes grind it to whitish flour between the thumb and index finger. Such a "soft" crystal may be beautifully green and completely transparent, but it is not solid. The emerald must be dried carefully and gradually in a sealed box for a long time to keep it reasonably clear and free of cracks. For this reason, there are so extraordinarily few good and flawless emeralds.

The *chrysoberyl*, as well as euclase and phenakite, occupy a special position between these two groups. They are found together with the emerald and are found with other beryls within the pegmatite cavities. The alexandrite, in particular, represents in its green-red color duality in the mineral realm something that only comes back in the higher stages of human beings and animals in the duality of venous and arterial blood. This gemstone also makes one think of the green-red fluorescence of the pure plant pigment chlorophyll when dissolved in ether.

Alexandrite is one of our rarest gemstones and the only one that shows the described double coloration in different lighting conditions. It got its name because it was found for the first time in the ancient emerald mines at Takovaya around the middle of the last century on the day Alexander, the Russian heir to the throne, reached adulthood.

As we mentioned above, beryls can form huge crystals. Although these large forms are mostly opaque and cloudy, they have no gemstone character; however, there are also very large crystals of, for example, aquamarine. In the province of Minas Geraes in Brazil, an aquamarine crystal was found in 1910 that was 48.5 centimeters long and weighed 110 kilograms. Besides Brazil, aquamarine is found in the Ural Mountains, in the Mountains of Mourne in Ireland, in the Namib Desert in Southwest Africa, and water-clear crystals on the island Elba. From Madagascar come especially pure-blue aquamarines and the fiery-reddish morganite. Also, California is a place where beautiful Morganites are found. Heliodor also comes from Southwest Africa.

Of the emerald deposits, the oldest is at El Qoseir (Kosseir) in Upper Egypt, which Sesostris already exploited around 1650 BC. All of the emeralds found in Egyptian mummy graves as decoration and perhaps also those legendary giant crystals of about 6 meters in length, which Herodotus describes in a temple there, came from this place. Moreover, the emeralds of old church treasures from the 8th century may have come from this place, as well as the stone that was found in the tiara of the Pope from the time of Julius the Second.

Among the older emerald deposits is Hohe Tauern, Austria, near Habach in the so-called Legbachgraben, a side valley of the Habach stream, which flows into the Salzach. The emeralds lie there in a mica-schist at the height of over 2000 meters above sea level on the edge of snow and ice. The ancient Romans and the Phoenicians are said to have gathered emeralds there.

It was not until 1831 that farmers digging tree roots found the famous Takovaya deposit in the Urals. Those places must have been known to the ancient Scythians, whose emeralds are reported by Pliny in his "Natural History." One of the largest crystals found there was in the Russian Imperial Mining Corps collection, and it was measured at 24 centimeters long by 15 centimeters thick. A similar giant crystal (emerald!) was also worshipped by the ancient Peruvians or the Incas, who were deprived of their treasures after the conquest of Peru. This stone is said to have had the size of an ostrich egg and came from the then famous mines of Muzo in Columbia, which are still exploited today.

In 1927 another deposit was found in North Transvaal in South Africa. Some beautiful emeralds also come from North America, but like the yellow Brazilian ones, they do not equal the quality of the older deposits from the Urals and Columbia. Also, the Australian and Indian deposits are without importance.

The common beryl, which has no gemstone value, occurs with the giant crystals in some of the places already mentioned. The largest of these is the crystal found in the Bumpus Quarry near Albany, Maine, which is 6 meters long, 1 meter thick, and weighs 18 tons. Other crystals found in New Hampshire weighed up to 1500 kilograms. At Limoges in central France, crystals as thick as an arm were found. In the Spanish province of Galicia at Ponferrada, six-sided beryl columns were used as doorposts.

Beryls crystallize in six-sided prisms (columns) terminated on one or two sides by flattened pyramidal end faces. Combination and intergrowth can produce nearly round shapes distinguished by intricate, polyhedral end faces.

Phenakite and euclase have somewhat different crystal forms.

Chrysoberyl and alexandrite come from the already mentioned Takovaya in the Urals and Sri Lanka. In Sri Lanka and Brazil, the chrysoberyl cat's eyes (cymophane) are found.

The crystal forms of the chrysoberyl are rhombic columns in various complicated arrangements. With alexandrite, star-shaped triplets occur by the crosswise intergrowth of three crystals. This strange gemstone shows by its crystal form that it is unique.

## 12. Tourmaline and Spodumene (Kunzite, Hiddenite)

In the same rock regions of the pegmatites, where we first met the rose quartz, topaz, and beryl, we also find the tourmaline and spodumene. Among the variety of minerals that line the cavities of these pegmatites, *tourmaline* is probably most remarkable for its extremely complicated material composition and its extraordinary variety of color.

Except for the garnet, the tourmaline is the most complicated gemstone we know. We certainly do not want to interpret the essence of a gemstone from its material composition, but if one finds with the tourmaline that no crystal is materially the same as the adjacent one, and if one further observes how the colors appear and are distributed playfully once like this and again like that on and in a crystal, then one is immediately tempted to think of a hidden artist's hand that has created these formations with inexhaustible fantasy.

Obviously, this artist's hand is still active today in the world of flower formation. We have already seen the colorfulness of plant blossoms shine from ancient times as an image of the finest, illuminated states of metals and similar materials. Yet, hardly any other gemstone reveals so much of this past living world of interweaving minerals, plants, and the butterflies' most delicate insect life as the tourmaline. It is, so to speak, the most "organic" of the gemstones.

Before we continue to the material composition of tourmaline, I will attempt to describe the almost indescribable colorfulness of this gemstone.

Some crystals display a more or less red color at one end and a tender green at the other end. This color contrast of red and green is characteristic of the tourmaline, and it can appear in all gradations of these colors. In between, there can also be irregular colorless zones. The distribution of the colors is rarely completely even. Furthermore, crystals exist that are completely opaque from one side because a dark rind covers the three- or nine-sided column in longitudinal ribbing or striations. If one cuts slices of a few millimeters thickness from such an opaque crystal, the cross-sections often reveal a colorfulness that was not suspected from the outside. The core of the crystal can appear in all shades of pure red, and around this red core lies a concentric colorless zone, followed again by a green layer.

However, on the very outside, where the "opaque" rind is located, a magnifying glass discovers another colorless zone only fractions of a millimeter wide that follows the green layer, which gradually turns into a deep indigo blue layer and is only opaquely black at the edge. The color sequence can also be reversed, with a green nucleus surrounded by red and corresponding intermediate zones. The colors of the core and the enveloping layers can also be directly linked. Such colored crystals are found in Brazil, California, and Madagascar. There are pale green or colorless crystals on the island Elba, some of which gradually turn deep black or reddish towards the top. Conversely, black crystals with white ends can be found. One Brazilian tourmaline has white crystals with red ends.

These tourmaline crystals are enveloped in color and capture only a small part of the color spectrum. Among the more or less "monochromatic" tourmalines, there can be every color shade imaginable. The abundance of flower colors spread over numerous plant families has been gathered and is sealed in this gemstone like a preview of possibilities from the ancient world. There are colorless, water-bright crystals called *achroite*, with a faint hint of red or green. These tourmalines are called *rubellite* (apyrite) if this hint of red densifies into peach blossom. A very common color is the brown of *dravite*, which ranges from barely transparent dark brown through all the gradations to yellow, reddish-yellow, and reddish-brown.

Very rare is the blue tourmaline or *indigolite*. It is most appreciated in the pure indigo tone and also occurs in cobalt blue, black-blue, greenish, and in the colors of sapphire and aquamarine.

Green is the most common color of tourmaline. Its range goes from yellow-green to blue-green, sometimes lighter, sometimes darker. Crystals with a chrysolite-like yellow-green to yellow and honey-yellow come from Sri Lanka. Emerald green is very rare and valuable. The black tourmaline or schörl is common and used as mourning jewelry. It often forms in certain granites on the surfaces of joints as radiating "tourmaline suns." In the mica rocks of the Ziller Valley in Tyrol, 25 meters (!) long schörl crystals have been found.

The through and through red tourmaline, the so-called *rubellite*, or *pink elbaite* (daourite), are rare and valuable. These gemstones exhibit all tones from the lightest to the darkest red, even the rare pigeon's blood color of rubies and the violet-red of amethyst. Not all possibilities have been exhausted with what has been described so far. In addition to the red-green contrasts already mentioned, multicolored tourmalines may also unite three or more colors.

As with other gemstones, the variety of colors is caused by the fine distribution of metals. Above all, iron, manganese, chromium, titanium, and iron-related cobalt in varying amounts and various combinations produce an astonishing abundance of colors. In addition, there is *lithium*, a substance related to potassium and sodium that does not by itself produce color, but in combination with the subtle coloring of manganese, it results in wonderful bright stones with red, light red, peach blossom, and light green. The precious rubellites are also so-called manganese-lithium tourmalines. *Iron* produces rich green and blue tourmalines; chromium gives an emerald-like tone, and cobalt that tinge of violet that distinguishes some rubellites. Cobalt is in all blue-green and blue tourmalines.

Figure 25: Elbaite tourmaline, Santa Rosa Mine, Minas Gerais, Brazil
(Photo by Rob Lavinsky, Courtesy of Wikimedia Commons)[xix]

As you can see, the same metals are coloring the other gemstones. The special thing about tourmaline is that the metals and their distribution are extraordinarily variable in multicolored crystals. In crystals having various sheaths and cores, there is still a certain order in the concentric color sequence that runs from the inside to the outside. Nevertheless, it is difficult to understand that such "ordered" formations consisting of materially quite different "layers" can only be formed by chemical processes. This "layering sequence" rather reminds one of the formation principles of the agates. It also reminds one of the layered construction of some plant stalks, where differently colored and formed cell layers produce similar cross-sectional images.

If one tries to imagine how such a “layered” tourmaline crystal could have been formed, the phenomena force one to assume that the varying metallic color clouds washed over and blew around the developing crystal because the basic substance of the tourmaline is a colorless material, which can be understood chemically as a complicated alkali-calcium-magnesium-alumina-silicate. In addition, there is also *boron*, which can form glass just like the silica and *fluorine* that we know from topaz—however, all these substances in combination form only colorless crystals.

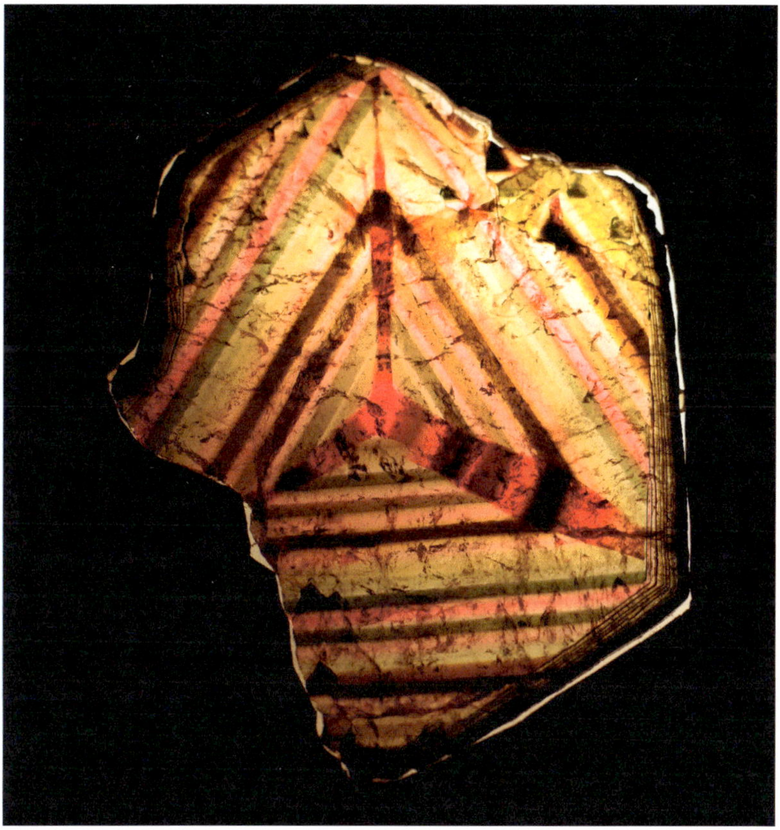

Figure 26: Multicolored tourmaline
(Photo by Sailko, Courtesy of Wikimedia Commons)[xx]

Like all the gemstones we have considered, the crystallizing tourmaline substance was previously a gel. The metallic element migrated rhythmically into the gel in the sheathed crystals, like successive "seasons." In the case of crystals that are unevenly "cloudy" at the bottom and top, the changing "metallic clouds" have penetrated the colorless gel mass from "above" and "below." Therefore, they display colorless but cloudy zones.

More than twenty different "elements" have been found in tourmaline crystals. Even the already mentioned basic colorless substances are very different at the same location. Also, there is the play of metals that causes colorfulness. No other gemstone displays such a lack of uniformity in its material composition.

This phenomenon expresses something that belongs to tourmaline's overall nature: *the strong laws of substance and form that most other gemstones reveal in their sharply outlined crystalline forms or exhibit in their gel structure are only partially realized in the tourmaline*.

The tourmaline's nature can be recognized because the mostly columnar formations build real crystal planes only at the ends. The long-sided "planes" of the column, which can be 3 to 9 sided, are not true planes but always slightly arched and formed with fine longitudinal ribbing. This longitudinal ribbing is a faithful copy of the trait we find widespread in the plant realm as the ribbing of plant stems. It occurs very clearly in most umbelliferous plants (for example, lovage, angelica, etc.) and many grass types. Here, this ribbing is nothing other than a filamentous or radiating development of the substance. We have already learned about it earlier with the greenstones and the description of the asbestos and similar formations in the mineral realm that do not achieve an actual crystal form.

Wherever we encounter this fibrousness and radiating quality of substances in the mineral realm, we must speak of a plant-forming tendency to overcome the lawfulness of the mineral. In this sense, tourmalines are the most transparent gemstones and the least crystalline mineral. Within the process of matter becoming transparent, a radiating-stem-like quality is interwoven with a colorful flower-like quality.

However, the stem and flower formation of the plant is something through which the sun's effect on the earth is expressed. It is the *sun* that, with its rays, draws the stem of the plant out of the earth, lifting it into the realm where the color-giving forces of light can wash over it. The sun that made the tourmaline crystals "mature" and then penetrates them with the rainbow of metallic colors was still *within the earth*. When the sun separated from the earth, the "tourmaline plants" died, and the subsequent world of flowering plants could develop with the sun outside of the earth.

The Dutch brought tourmaline from Sri Lanka at the beginning of the 18th century, known by the Sinhalese name *turamali*. This word's meaning is "red stone;" it was also used for garnets and zircon. In ancient times, gemstones were not distinguished by their material properties as they are today, but only by their *color*. The corresponding Medieval-Central European word for red stone would be *carbuncle*. The average finger-long and finger-thick crystals of the tourmaline have still another special characteristic which the Dutch already noticed: by warming and then rubbing it with wool or silk, the crystals become electrically excited and attract small pieces of paper, similar to amber. It was called an "ash extractor" (ash puller) because the Dutch used it to pull the ash out of their flue pipes. These electrical properties are related to the crystal structure of tourmaline: when there are *more* surfaces at an end, positive electricity occurs, negative electricity occurs at the end with *less* surface.

Despite its relatively low silica content of 32 to 38 percent, tourmaline is usually somewhat harder than rock crystal.

The locations of these gemstones are mainly on the American continent, South Africa, Madagascar, and the Urals. In North America, they are mainly in California, Colorado, Maine,

Massachusetts, Connecticut, and New York, where tourmaline-bearing pegmatites are located in the mountains.

To list all sites in detail would be too much. Remarkable are some California sites, where crystals have a length of up to 30 cm long and an average thickness of 10 cm. Also, quite large crystals are found in Madagascar. Here, as at the Brazilian sites, all colors and color combinations occur. From Africa in the Namibian Desert, blue-green and blue crystals originate that become brighter and fiery when heated to about 650 degrees. There are rich sites known in remote areas that cannot be exploited because of difficult mining conditions. In the Urals, good but smaller tourmalines of unusual colors are found in many places. They occur together at many sites mentioned above with topaz and beryl.

In some California sites and those in Madagascar and Brazil where tourmaline is found, the more precious mineral *spodumene* is also found. Spodumene is a lithium-alumina-silicate and is not mineralogically related to tourmaline. The lithium substance has stronger relations to the silica and therefore occurs almost exclusively in mica, the so-called lithium-mica or *lepidolite*. In many locations on earth, the spodumene forms enormous crystals in pegmatites of up to 90 tons. But these crystals are not transparent and cannot be used as gemstones.

When the precious form of this mineral is pale lilac or violet, it is called *kunzite*, and when it is greenish to emerald green, it is called *hiddenite*. However, pink- and yellow-colored crystals also occur. The lilac to violet tones come from manganese content, the yellow from iron, and the green from chromium. So we find the same metals again as in the other gemstones.

Green Hiddenite takes its name from its discoverer. W. E. Hidden first found this gemstone in 1881 at Stony Point, North Carolina, and described it as a lithium mineral. It was previously thought to be diopside, a green stone mineral that we have already discussed. At this locality, it is accompanied by tourmaline and other gemstones.

The much more common lilac to violet *kunzite* takes its name from the mineralogist George Frederick Kunz. It is found along with tourmaline at some mines in California, Maine, Madagascar, and Brazil. It was discovered in one of the California mines in 1902.

These precious spodumenes form remarkable columnar but flat crystals that are tabularly cleavable along the longitudinal axis. In the more precious form, they display excellent luster and great clarity with delicate colors. The hiddenite crystals are generally smaller than the kunzites.

The spodumene crystals seem to be composed of very fine platelets the length and width of the crystal. This is the reason for their delicate cleavage, which makes polishing difficult. This structure is called "spar." It occurs in many other minerals, such as feldspar (from which it gets its name), mica, and gypsum. However, with spodumene, it indicates this mineral occupies a certain intermediate position between mica and feldspar. After all, these are minerals that accompany spodumene in various forms.

Mineralogists often treat the spodumene together with the hornblende-like augites. However, these minerals do not occur together with the spodumene and appear later in the rock world. Therefore, one can consider the spodumene as a precursor of the hornblende and augite, which are much darker because they contain metals in an already denser form. There are no actual gemstones among them.

With the tourmaline, one has to do with a similar intermediate position, which could be characterized between feldspar and hornblende. Thus, from the feldspar, it has the interesting crystal end faces, and from the hornblende, the long-form with the finely ribbed sides. Since tourmaline contains 2 to 3 percent *water* (!!!), it approaches certain feldspar-related minerals. For this reason, mineralogists may treat it as one of these minerals, even though it does not fit into any of the existing classifications.

All these phenomena associated with tourmaline and spodumene demonstrate that one has to do with extraordinary formations. They are unique creations, and they proclaim very clearly something of the strange primordial life of the earth, about which we have spoken repeatedly.

## 13. The Precious Corundum (Ruby and Sapphire), Spinel and Zircon

A very peculiar rock community is found at a few places on earth. In such places, the frequently mentioned *pegmatite dikes* and veins from previous chapters meet with very old *crystalline schists* and marble-like *limestones* or *dolomites*. The crystalline schists have a special relation to the greenstones mentioned earlier, which were characterized as an expression of a tree-forming tendency in the ancient life states of the earth.

The third component of this association, the marble-like limestones and dolomites, points to a new impulse in the earth's life history. This needs to be looked at briefly to understand that *corundum* appears within the rock world just at those places characterized above.

The limestone in the Jura Mountains, for example, is a definite formation in more recent layers of the earth and contains innumerable sea animals which left behind their casings and shells after they died. Where deposits of dolomite (which is often permeated with silica) appear between these limestones, it can be assumed that certain lower plant forms dominated sea life. These lower plant forms are algae, like the ones we still know today. Single-celled or multicellular algae species that secrete lime or dolomite in their life processes are still widespread.

Dolomite is a mixture of calcium carbonate and *magnesium* carbonate and suggests its plant origin by its magnesium content. As the kernel of chlorophyll, magnesium is *the* central substance for the functioning of the plant world's life processes.

Lower plant forms, like the algae, can deal with calcium and magnesium, similar to the way lower animals with their casings and shells do later. This indicates the presence of an *animal impulse* in these early forms of the plant realm. A clean separation between animal formation and plant formation does not yet exist. This "animal impulse" is still very clearly found with the ferns in their complicated way of reproducing. We cannot present more about the strange intermediate position of lower plant forms because it would cause us to diverge from the task at hand. I refer all interested in the subject to the excellent work of Görbert Grohmann in his book "The Plant" Vol. 1, 1989, Biodynamic Farming & Gardening Association.

For these oldest times of the earth's life, one must now consider the marble limestones and dolomites, which, as described above, are embedded between pegmatites and crystalline schists, as formations from life processes of lower *plant* forms. These so-called primeval limestones and dolomites have nothing at all to do with the later animal limestone deposits.

Let's now look at the totality of what is expressed in this "rock community" of pegmatites, crystalline schists, limestones, and dolomites. We could characterize it in the following way: the *pegmatites* with their cavity formations point to the beginning of the separation of the animal-forming principle from the old mineral-plant world of the granite, which we have already met earlier as a world of flowers proliferating into each other. But, on the other hand, the gemstones, which appear in the pegmatites as individual formations, should be regarded as detached blossoms: small single minerals hidden in the granite rising to the full beauty of their crystal forms.

In the *crystalline schists*, the mica's leaf principle intensifies to the described "tree formation," expressed very clearly in the actual greenstones. The "plant" thus appears in a form whose metamorphosis becomes the trees of today. In the realm of schists and the characteristic greenstone formations, we don't find any indication of "animal impulses."

Instead, embedded in the limestones and dolomites, the "animal impulse" of the feldspar now appears in the granite, as if detached and separated and completely transformed. For in the trinity of quartz, mica, and feldspar in granite, the feldspar is the "calcareous" component. As an individual component of these rocks, the "calcareous" disappears completely in the crystalline schists. But it appears, as said, new and transformed as the marbles and dolomites, which are interbedded with the schists.

The still quite undifferentiated mineral-plant life of the granites divides and develops itself so that first, in the pegmatites emerging from the granite, the *flower principle* begins to separate (gemstones). Next, in the *schists*, the *leaf principle* makes itself independent and thickens into the *stem-like*. Finally, a further new life impulse shows up in the marble-like limestones and dolomites, pointing to the *lowest plant forms* that still have animal properties in their life processes.

One must imagine now that the gel-like condition of these later rocks originate successively from the described life processes and are at first still existing plastically next to each other. All those peculiar penetrations, superpositions, and effusion-like movements come about, which one can then observe in the solid rock. The magnificent folds in schist and gneiss are not conceivable without this plastic condition.

At this "point in time," which brings about the new life effects described, *corundum* is formed, whose precious forms have been known since ancient times as *ruby* and *sapphire*.

The basic substance of all corundum is crystallized *alumina* (aluminum oxide). We have already encountered this alumina in the precious feldspars, in the beryl and the topaz, connected with silica and other substances. With the corundum, we deal with "pure" alumina, which, as with all other gemstones, is colored by metals.

Alumina is also found in mica in combination with silica and potassium or magnesium. In addition, it is found in the feldspars together with silica, potassium, sodium, or calcium carbonate. These two minerals, mica and the feldspars, account for over 60 percent of the rock structure. Thus, it is possible to see what role the alumina played in the earlier stages of earth life.

However, there are still plants today that process alumina to a significant extent in their structure. To these belong the horsetails, the ferns, the moss ferns (Selaginellaceae), and some Proteaceae that occur only in Australia. In the life processes of these plants, the alumina has special relationships with leaf formation. In the mineral world, or rather in the earlier life realms of the earth on which it is based, this leaf relationship is shown in the formation of mica, which, as we mentioned earlier, goes back to sepal leaf-like formations of the ancient mineral-plant world.

Therefore, in the mica, the alumina, together with the silica, is the substance that determines the characteristic leafy structure.

The remarkable thing here is that a substance, which is very common in combination with silica, appears independently and separately. This is a similar phenomenon with other gemstones, where common substances also appear in a precious form, with the significant difference that *free* alumina is a great rarity. The appearance of this *free* alumina as corundum is a phenomenon of differentiating life processes just as the emergence of the shists and limestones.

In the detachment of the alumina substance from its close association with the ancient mineral-plants, world forces are expressed whose uniqueness is only surpassed by those underlying the formation of the diamond. It must be noted that a substance commonly found in abundance in the soil of every farm field is here lifted out and refined like almost no other earth substance.

Rudolf Steiner said that the origin of these world forces is to be sought beyond the zodiac. We do not presume here to get behind these secrets.

This free, crystallized alumina substance is exceptional because it is not bound in its occurrence to *one* of the three rock types mentioned, but it can be found embedded in the pegmatites, schists, limestones, or dolomites. Furthermore, since the crystalline schists have a strong evolutionary relationship with the greenstones, it can even happen that, for example, sapphire is also found in basalt, which is related to the greenstones.

Another special feature of corundum is its *hardness*. According to the already mentioned numerical Mohs scale, which does not reflect an absolute value, the corundum has a hardness of 9. In reality, the corundum is about nine times harder than the rock crystal, and in this property, it is surpassed only by the diamond, which is ninety times harder than the corundum.

There are many crystal forms of the corundum. *Rubies* are mainly six-sided prisms with flat end faces, and *sapphires* often have six-sided barrel shapes that end flat or in six-sided pyramids. The transparent red corundum has always been called *ruby*; the so-called pigeon's blood color is rare and much appreciated. The ruby's coloration results from an iron content and a trace of chromium so minute that it can no longer be detected analytically but only by spectrographic methods.

Blue corundum is called *sapphire*. Its coloration is also based on iron with a trace of titanium, which is related to iron. White and yellow corundum is also called sapphire because it has been found that the coloration is based on the same metals and varies only by the size and distribution of the colloidal metal particles. All other colors, which occur in great variety and are found together with ruby and sapphire, are also based on these three metals. Some colored corundum varieties are given the name of other gemstones with the prefix "oriental." Thus, violet corundum is called "oriental amethyst," an emerald green corundum, "oriental emerald," topaz-colored corundum, the "oriental topaz," and so on. An exception is the reddish-yellow corundum "King's Topaz" or "Padparadscha" from Sri Lanka.

The asterismus is an optical phenomenon occurring in rubies and sapphires and rarely in the garnet to be discussed later. In many faint or opaque colored stones that cannot be cut with facets, this phenomenon manifests as a star-shaped six-rayed glow that moves according to the light source. These *star rubies* and *star sapphires* are cut in a spherical to hemispherical shape to emphasize their appearance. This movable light star phenomenon is due to the crystal structure of corundum, in which the finest crystal needles of other minerals are embedded. Such stones are often only dove-gray colored and have a dense "silk," that is, they are somewhat "foggy." Strongly colored stones with a star are among the great treasures and rarities.

Before we speak about the sites where precious corundum is found, above all the ruby and sapphire, I would like to refer to the explanations in chapter "9. Lapis Lazuli and Turquoise" regarding the old Lemurian continent. What was said there about the terrestrial periphery of this continent applies especially to the precious corundum. The main sites for discovering these precious gemstones are South Asia, Sri Lanka, Madagascar, and Australia. African corundum is known, but no crystals worthy of cutting have been found. On the South American continent, only a few corundum stones have been found in Central Brazil. North Carolina and Montana in North America have some sites that provide the ruby and the interesting Montana sapphires in addition to technical grade corundum. However, for centuries the classical areas for finding precious corundum have been South Asia and Sri Lanka. For ruby, the district of Mogok in Upper Myanmar should be mentioned first. The gemstone is extracted from the weathered mountain debris, as is the case in almost all localities. However, the parent rock of the ruby is known to be a coarse white marble, meaning the crystals are formed floating in the lime gel before the solidification of the marble.

Only a few people have seen the magical picture of a red ruby crystal formed in white marble because the gemstone material is obtained exclusively from the weathered rubble, debris, and clay. In addition to the ruby, blue, green, yellow, and white sapphires are found there, some of which came from neighboring pegmatites and are included in the rubble. The stones are very small, most of them weigh only fractions of a gram, and larger pieces of 6 to 9 carats without flaws (1 carat = 0.2 gram) is a rarity. And yet, every few years, stones between 30 and 400 carats are found worthy of cutting. Even rarer are large pieces of up to 1000 carats that cannot be used as gemstones. The Myanmar rubies are considered the most beautiful.

Other deposits are located in Thailand with similar rock conditions. Here, too, the stones are extracted from weathered debris by "placer installations." The Thai rubies have colors exhibiting a hint of orange-yellow and brown, so they are less valuable. Sapphires are also found in tandem here.

In Afghanistan, ruby-bearing marbles east of Kabul and elsewhere have been known for centuries. The yield in the past from these localities is said to have been exceptionally rich.

In Sri Lanka, light raspberry-red rubies can be sparingly found alongside abundant sapphires in placer installations in the island's southwest. Ruby is even rarer in Australia and is sometimes found together with sapphire in New South Wales.

The sapphire is found mainly in Thailand, Kashmir, and Sri Lanka. The Thailand sapphires are only occasionally found with rubies, and their parent rock is not exactly known. Since the rock community always has the described characteristics at all discovery sites, it is difficult to say anything definitive about the parent rock. Experience teaches us that the sapphire is bound to the pegmatites, schists, or even the greenstone-like rocks. In Thailand, as in the famous Kashmir, sapphires are also extracted from placers. This locality, which became known in Europe only since the eighteen-eighties, has become famous for its sky-blue and cornflower blue stones, which retain their magnificent color even in artificial light. The largest sapphires are found In Kashmir. Pieces 30 centimeters long have been found there, and crystals 15 centimeters long by 9 centimeters thick have been cut.

Most sapphires come from Sri Lanka. They are lighter in color than those from India and Thailand, have many silky inclusions, and change color with the light. Sapphires from Sri Lanka can be a dingy blue-gray in artificial light but then turn the most beautiful cornflower blue on a foggy or overcast day. The luster of Sri Lanka sapphires is more vivid. Large stones and those with six-rayed stars occur more often there. Variously colored corundum comes mainly from this island and is found together with sapphire.

Australia mainly supplies magnificent green and green-blue sapphires, whose parent rock is basalt. The stones are also obtained from placers like India and Sri Lanka but are characterized by greater hardness.

The sapphires of Montana in America are mined, but placers are also used. The stones quarried from the solid rock (a type of basalt) are characterized by more intense colors than those from the placers. Again, the hardness is greater than what is found at the South Asian sites. Among the Montana sapphires, there are also brownish shades with a metallic-blue reflection. The blue and green stones from there are famous, which, like the alexandrite, turn red in artificial light.

Closely related to the occurrence of precious corundum is the *spinel*. These gemstones are lesser-known because they are identified by other trade names, such as balas ruby or alabandine ruby. Spinel, however, is materially something quite different. It consists of a mixture of alumina, as it is in corundum, and magnesium oxide. Its crystal form is octahedral (two four-sided pyramids placed on top of each other) with an inclination to form a rhombic dodecahedron. It is found almost exclusively together with the ruby and sapphire in placers. However, its parent rock is always marble or dolomite, in which it was floating when it formed.

The spinel should be regarded in the same sense as corundum as an extraordinary form and is close to corundum developmentally (but not materially!). The magnesium content of the spinel points to a colorful "flowering" lifted from the mica-leaf-like formation, even more so than the corundum. Red, blue, violet, brown, and yellow-green are the essential colors of this gemstone and are dependent on the same metals as the corundum. The discovery sites are the same as for ruby and sapphire; emphasized are Myanmar, Thailand, and Sri Lanka. A yellow spinel comes from Brazil; North America has some sites which occasionally yield stones worthy of cutting.

The hardness of spinel is less than that of corundum, being about the same as topaz (hardness 8).

Unlike corundum and spinel, *zircon* is a common mineral found throughout the world. It is found in many rocks, and there is hardly any river or sea sand that does not contain zircon, at least as microscopic grains. *Only where zircon appears in the company of precious corundum does it become a gemstone.* This fact impressively illuminates the constellation of gemstone forces because it appears at the meeting of the three types of rocks described above.

Zircon is the siliceous compound of a remarkable metal of the same name, which is extraordinarily durable and as difficult to melt as iron. Moreover, the metal obtained from zircon has a clear relationship to the metal obtained from silica, the so-called silicon.

The metals from silica and zircon are as hard as quartz, can carve glass, and, alloyed with iron, confer great hardness and acid resistance.

Clearly, the zircon substance has a great relationship with silica, with which it is always associated. It can be considered a highly densified silica, considering it has almost twice the latter's specific gravity. Thus, zircon is the *heaviest gemstone* known. This relation to gravity is expressed in another way: *every zircon contains radioactive substances.* It is the thorium that makes zircon the main carrier of radioactivity in rocks. Thus, all radioactive substances such as uranium, radium, and so on belong to the rock world's heaviest, most densified substances.

Zircon crystals are four-sided prisms terminated on both sides by four-sided or rhombic pyramids. Its hardness is slightly higher than that of rock crystal, specifically 7.5.

The refraction and brilliance of zircon are so strong that they can be easily mistaken for diamond. This brilliance is also the reason why its multiple colors appear extraordinarily fiery. This quality is especially true of the most precious zircon form, the fiery red hyacinth, which can go into the orange-brownish. Other hues include: rose red, brown, yellow, and purple. Green in various gradations is not uncommon in Sri Lanka, while blue is produced by firing the brown stones. Unfortunately, this firing is also applied to bright hyacinth, resulting in colorless diamond-like stones traded as Matara diamonds (after Matara, where it is found in Sri Lanka).

The coloring metals are the same as with corundum and spinel, except the blue zircon of Thailand contains some copper and the rare element hafnium, in addition to iron and titanium.

Zircon, where it occurs in the rock, is so intimately bound that it cannot be extracted. Therefore, it is found only in the already mentioned gemstone placer installations together with other gemstones. The principal sites are in Sri Lanka, where the largest quantity and most beautiful zircons come from, and Australia and India, where corundum is found. In South Africa and Brazil, hyacinth accompanies diamonds.

## 14. The Diamond

When considering the diamond, we enter a place that is even more hidden for today's human cognitive faculty than everything connected with the other gemstones. This is already evident from the simple fact that with the diamond, we no longer deal with an ordinary inorganic mineral substance, such as silica, alumina, or their compounds, but with *pure crystallized carbon*.

Carbon is the substance that transforms invisible life into a *visible form* and is the basic scaffolding of *visible* life on earth. Carbon is also connected with the shape- and form-giving properties that appear to us in its everyday form as graphite, bituminous coal, and lignite, particularly as the remnants of former life in a dark, opaque substance. The blackness of carbon is an expression of its inherent powers of densification and shaping. Life itself has created this substance to bring its flowing formations, which originated entirely in the invisible realm, to visibility. Thus, *a substance was uniquely created out of life, unites the world's polarity of light and darkness,* and connects the upper and the lower, the invisible and the visible.

In this sense, carbon is *the* substance born out of light, which at the same time saturates itself with the highest density of earth; it is the most light-permeable substance and the hardest substance we know.

Carbon appears in the earth's rocks in three different forms: *diamond, graphite,* and *coal* (bituminous coal). The layers and formations in which these three forms of carbon are contained indicate that they belong to completely different times in the earth's life.

The *diamond* appears in South Africa embedded in a rock closely related to the already mentioned greenstones: kimberlite or "blue ground."

On the other hand, *graphite* has its home mainly in the schists of older strata, in the associated marble limestones, and to some extent in the pegmatite dikes and veins of granite.

It is commonly known that *coal,* as bituminous coal, builds its own formation. Between graphite and bituminous coal, especially anthracite, there are segues to certain older coal deposits.

Rudolf Steiner pointed out that these different forms of carbon are connected with the older states of existence preceding the present state of the earth. Rudolf Steiner described these "planetary" states of the earth in detail in his book "An Outline of Occult Science" and called them the Saturn, Sun, and Moon stages. Since it is not possible to recapitulate the essence of these representations by Rudolf Steiner in a few sentences, it is necessary here to refer expressly to this work, which forms the basis of this book "Gemstones and Earth's History."

The "*Saturn stage*" is a state of pure warmth and plainly the beginning of the world of "matter," but as yet showing nothing of "life."

The "*Sun stage*" is a world body of warmth, light, and air, in which a "plant-like" life first appears.

During the "*Moon stage,*" the "fluid" and the first predispositions of "animal-like" life appear.

Only on the present earth, in the fourth stage, does the solid-state appear. The human being also appears here in its bodily form after passing through all the earlier developmental stages.

At the beginning of the *earth's* present state, the past states of Saturn, Sun, and Moon are *repeated* and form in the *rock world* as granular (siliceous) rock, layered (schistose) rock, and limestone.

Therefore, we have the structures and the substance of the rocks as a key for the origin of these formations from the earlier stages of the earth. Where these formations interpenetrate, where their structures and materials blur and mix, we must remember the *recapitulation* of the past in the present stage of the earth's development. We should not be surprised that the limestone already appears with the granites and gneisses because the "animal-impulse" connected to it, which already arose during the "Moon stage;" it already began to make itself felt on the *earth* when the "Saturn stage" is repeated. This animal-impulse pushes itself forward and becomes stronger and stronger until it has the upper hand in the lime layers of the Jurassic and the repetition of the "Moon stage." Conversely, "granular rocks" can break through in the limestone of the Jurassic as so-called volcanoes because the past "Saturn state" is still alive and active. These seemingly confusing phenomena cannot blur the fundamental law of rock formation present in the described threefoldness and can be seen by a person of goodwill.

The *diamond* appears in certain *greenstones* in company with *garnet* and *chrysolite* (gem-quality olivine). As we will see later, the garnet and chrysolite can also appear separately and in completely different rocks. But this is connected with the fact that the diamond's occurrence is bound only to quite specific areas of the earth. They are the same areas we already learned about with the turquoise, lapis lazuli, and corundum. In any case, garnet and chrysolite are always connected with the greenstones or cross over (especially garnet) into the shists that are related to the greenstones.

These greenstones in which the diamond appears stand in a special relationship to the "Sun stage" of the earth and reveal the "time" when the sun, moon, and earth were still *one* world body. Here we have to remember that "plant life" came in during the "Sun stage," which then expresses itself through earthly rock development in the characterized properties of the greenstone minerals. Therefore, one must imagine that the diamond originated during this "Sun stage" when the whole earth exhibits a strong "plant" life. But within the recapitulation of the Sun stage on earth, the Moon stage also makes itself known. This becomes visible where, in the gneisses and schists (now, however, far away from the greenstones), the lime (marble) inserts itself. *Where this* is the case, one finds in the gneiss, schist, and lime (marble) the *graphite,* which Rudolf Steiner calls that form of carbon originating during the earth's Moon stage.

The third *earth form of carbon*, the earth form, as Rudolf Steiner calls it, is bituminous coal, which only develops after the sun and the earth separate. Thus, at the beginning of this formation of bituminous coal, the "recapitulation of the Sun stage" is also completed, which expresses itself very clearly in the predominant schist formations.

In the *diamond*, we can see something "Sun-like," in the *graphite,* something "Moon-like," and in the *bituminous coal*, something truly "earth-like."

These three forms of carbon point to three different life stages. We still find perfect life forms preserved as shadowy imprints of the primitive plant forms in bituminous coal. Nobody doubts that this non-crystalline coal has emerged from very undifferentiated plant life.

With *graphite, such* life forms are not found. We see something else. Graphite is leaf-like and, in its basic form when it crystallizes, is very similar to the six-sided mica foliations. It occurs in rocks such as granite, gneiss, and schist, and there it "displaces" this mica and "takes its place." This means that where mica once existed, the graphite suddenly appears. If we remember that the mica is the "leaf element" of the ancient mineral-plant world, then we see that *this* form of carbon, the graphite, was related to "life." This is also shown in yet another way. There are graphite deposits that are fibrous, flakey, platey, and even wood-like masses are not uncommon in the large deposits of Sri Lanka. However, these are all forms we have already found earlier with the greenstone minerals and the precious ores. The life, whose "remnant" is the graphite, was very close to the old mineral-plant world. In the appearance of *this* form of carbon, we see the first germ-like appearance of later carriers of life and form. It is the first attempt by creation to transfer the old "silica world" into the subsequent "carbon world" of life.

The diamond appears in those rocks that exhibit the first *large* beginning towards organic wood-like formations in the mineral world, but the "organic" is mostly blurred. The substance of the diamond is pure carbon, thus an organic substance. But the rocks that contain this pure crystallized carbon are not to be regarded in the actual sense as typical carbonaceous rocks. There are innumerable worldwide occurrences of rocks similar to the diamond-containing "blue ground." However, they contain neither carbon nor graphite, from which, under certain conditions, the diamond could have "formed." Therefore, carbon is a stranger in these rocks if one wants to speak in the sense of modern geochemistry.

But in certain places on the earth, this "stranger" is nevertheless present as the remainder of the "Sun life" of the earth, a life whose "formations" are beyond our imagination today.

There are features of the diamond that let us presume something of these primal forms of life. The diamond's most frequent crystal form is the octahedron, that double four-sided pyramid we got to know as spinel, *the* sun-form among the Platonic crystal bodies. But this octahedral shape of the diamond crystal strives through innumerable variations to the *spherical form*. All the transitions from the octahedron to the rhombic dodecahedron with intermediate curvilinear forms up to the spherical form are found. These curvilinear forms have convex and concave "planes" and are absolutely characteristic of the diamond; we find them almost exclusively in the diamond. However, in fully spherical form, we find them only in those forms of the diamond which accompany it as the non-precious and opaque bort, ballas, or carbonado. *Bort* is a dense, black mass with a random or *radially-fibrous* internal structure and is often purely spherical. *Ballas* is a variety of bort made of concentrically arranged smaller diamond crystals with a *fibrous texture*. *Carbonado* is a massive, opaque pea- to egg-sized diamond with a shiny black surface. These ignoble diamond forms may be *harder* under certain circumstances than the pure, transparent crystal and, therefore, desirable material for deep drilling.

In the spherical forms of the diamond and the radially-fibrous bort spheres, one can recognize that the diamond's *original form*, when it was still in the gel state and before the crystallization forces seized it, was the *drop*, the *sphere*. The sphere, however, is the *archetypal form* of life.

Other minerals and gemstones can be embedded in diamond crystals. These crystals have been found enclosing garnet, chrysolite, topaz, quartz, and zircon crystals so that these other minerals can be clearly seen floating within the visible diamond crystal. Liquid inclusions are not too rare; they have been identified as salt solutions and liquid carbonic acid. These phenomena show clearly that the diamond must have originated from an aqueous gel state because pure carbon is infusible, and the included minerals all have melting points around 1500 degrees. The diamond may even have hardened much later than the enclosed minerals. Diamonds from Brazil are known to have become tabularly intertwined with rock crystal, a phenomenon that confirms the common gel state.

Major diamond locations have been discovered in Brazil and South Africa. They are the already mentioned serpentine greenstones, which are called kimberlite or blue ground. The famous "pipes" are known in these two places, which are "volcanic," tubular vents filled with kimberlite rock. In South Africa, the diamond is almost exclusively extracted from these pipes by blasting and crushing the rock. Due to the hardness of the diamond, hardly any stone is destroyed in this rough process. The crushed rock is mixed with water, and the slurry is run over inclined shaking tables that are thickly coated with grease. Only the diamonds remain stuck in this grease; all other minerals are spun away with the water. This adherence of the diamond to fat is a consequence of its organic substance (carbon), which has a surface tension quite different from other minerals.

The regional South African diamond deposits were discovered only in 1867. Children found the first stone, which weighed about 21 carats and was later displayed at the World's Fair in Paris. In the years that followed, diamonds were repeatedly found in various places in the Free State Province, including the famous "Star of South Africa," which weighed 85 carats. A systematic search began in the 1870s, leading first to the discovery of the "yellow ground," which was nothing other than weathered "blue ground," and in the following years to the exploration of the "pipes." About 250 of these were discovered over time, of which only 150 were diamond-bearing and only 25 worth mining. Finally, around the turn of the century, the systematic exploitation of these deposits began, which have remained the leading ones to this day and are considered practically inexhaustible. Depending on the location, 30 to 50 percent of the diamonds extracted are larger than one carat (0.2 grams). The largest diamond ever found to date is from the Premier mine's yellow ground. It was named the Cullinan after its place of discovery in 1905 and weighed 3,106 carats. Between the first discovery of South African diamonds and 1926, some 24 stones between 100 and 1600 carats were found.

The colors of South African diamonds range from prized blue-white and sapphire-blue to all shades of the rainbow, including red and pink. Yellowish tones are predominant. Diamond crystals that exhibit a dark, distinct cross in their interior are rarities. This form comes about because of the layered structure of the crystal, which contains darker substances. Due to its fourfold symmetry, the regular inclusions appear as a cross. Such a stone was given to the Red Cross as a gift from the South African diamond mines during the First World War.

In addition to these sites in the pipes, there are numerous placer installations for diamonds in Africa, where the gemstone is found loose in sand and weathering debris. Some of these are located in South Africa, Southwest Africa, Zimbabwe, the Tanganyika area, the Democratic Republic of the Congo, Angola, and Liberia.

In 1929, the world production of diamonds was about 7 million carats, of which Africa alone supplied 6.5 million.

The oldest sites for diamonds, which were exploited in pre-Christian times and are now exhausted, are in the Indian frontier on the eastern side of the Deccan highlands, which consist entirely of massive basalt covers. There the diamond was extracted from placer installations, conglomerates, and the debris of weathered rock. These Indian diamonds are characterized by their size and purity.

The famous "Orlov," which is possibly the "Great Mogul," is a pale blue-green stone of about 400 carats and is considered the best Indian diamond. It was found in 1680 and cut into a diamond of about 200 carats, which, after many wanderings and ownership changes, was given to Catherine II of Russia in 1772, who had it set in the Russian scepter. As such, it is now in the Russian state treasury. Several other Indian diamonds between 400 and about 40 carats have been given names after cutting, including the Koh-i-Noor (Mountain of Light). These stones are in state treasuries, in museums, and the treasuries of Asian nabobs. The diamond deposits of Borneo are practically exhausted. The stones were found there only in placer installations. A special feature in Borneo was corundum, into which small to microscopic diamond crystals were embedded.

Australia has diamond sites in New South Wales, Queensland, and other regions. Australian diamonds are considered to be tougher than stones from other sites.

On the South American continent, the most important sites are in Brazil. At first, the discovery of Brazilian diamonds in the eighteenth century supplanted the domination of India. Then, at the end of the nineteenth century, Brazil was overtaken by Africa.

These Brazilian sites are located in the provinces of Bahia and Minas Geraes. Here, too, even today, the gemstone is extracted from weathering debris by placer installations. Brazil produced an extraordinary wealth of beautiful stones during its 150-year diamond mining heyday. Some of the largest stones are worthy of mention: the "Star of the South," about 260 carats, which was cut into a diamond of 125 carats and is now owned by an Indian prince; the "Star of Minas," about 180 carats; and the "Southern Cross," a pink diamond of 118 carats, found in 1929. Other less important sites are located in South America in the Co-operative Republic of Guyana and Venezuela.

North America has several sites in the United States and Canada that have yielded few stones overall compared to Africa, Brazil, and India. These North American localities are only interesting because they show the diamond is primarily found in the greenstones or related rocks, similar to South Africa.

In Russia, diamonds have been found in the gold and platinum washes of the Urals. This occurrence, together with platinum, points again to the greenstones. Findings in Lapland and the Czech Republic (Bohemia) are closely related to the garnet, which accompanies the diamond at most of its primary sites. In the Czech Republic, there were only two crystals of a few tenths of a carat found in the garnet-bearing sands of Diazkovic.

From these different, widespread locations, it can be recognized that the *possibilities* for diamond formation were present in many places on earth. The diamond is always found in the same rocks: greenstones of various varieties, serpentines, eclogites. And there are always minerals related to chrysolite (olivine or peridot), and rarely is the garnet missing.

But the really important discovery sites are, as already indicated, in India, Africa, and Brazil, thus in the area of the ancient Lemurian "continent," which we already mentioned with the Lapis Lazuli and Turquoise. It is that area of the earth where, according to Rudolf Steiner's research, the human being transitioned from a more spiritual existence into the beginning of a physically visible corporality. This process lasted until far into the Atlantean period (Tertiary). Its origin goes back to the already mentioned "Sun stage" of the earth. Therefore, Rudolf Steiner speaks about the formation of the diamond at the time of the first beginnings of the human physical body. These "first beginnings" are considered the "germs" of the later *sense organs*. First among these is the *eye*, the sense organ that consists of light-permeable matter. This light-permeable *matter* of the eye has a deep inner connection with the diamond's light-permeable, hard matter. The *ability* to *actively* see, on the other hand, has its mineral counterpart in the *chrysolite* that accompanies the diamond.

## 15. Chrysolite and Garnet

The *chrysolite*, known mineralogically as gem-quality olivine or peridot, is a companion of the diamond only insofar as it occurs as a *component* of kimberlite. The olivine-chrysolite accompanying the diamond in the kimberlite does not have a gemstone character because its grains are too small and impure. Together with the diopside that was already mentioned with the precious greenstones, the ilmenite, red garnet (Cape Ruby, Pyrope), and serpentinized kimberlite, it forms the mother rock of the diamond.

Materially, the olivine-chrysolite (peridot) is a magnesium-iron-silicate with a very low nickel and manganese content. Its colors range from yellowish olive green to bottle green; yellowish and brownish tones also occur. The precious chrysolite is a unique rich golden green that only rarely appears in other multicolored gemstones. Its hardness is slightly below that of rock crystal, so it is susceptible to damage when worn as a ring stone. Due to the color's uniqueness, this gemstone, which has become rare, is highly prized, and good-quality stones are difficult to obtain.

At the end of the chapter on precious greenstones, we pointed out chrysolite's unique position as the only gemstone found in *meteorites*. These meteoric chrysolites, such as those found in the famous Krasnoyarsk meteorite discovered in 1749 in Krasnoyarsk Krai, Russia, can be so large and well crystallized that they can be cut. Thus, the person who wears such a stone can say that they wear a gemstone that has fallen from the sky.

A great secret is revealed about the earth's formation in past times through olivine and meteoric chrysolite stones. We pointed out that the diamond originates from a life stage of the earth's development that Rudolf Steiner calls the "Sun stage." The chapter about the precious greenstones describes how mineral-plant life is closely connected with this "Sun stage." All these strange rock and mineral formations of greenstones, from the gabbro and serpentine to the basalt, which are found intercalated between granite, gneiss, and schist, or, like the basalt, break through the dried-up shell of the earth, have originated in this "Sun stage" of the earth's life.

Remarkably, *meteoritic rock is characteristic of the greenstones* already mentioned. Many typical minerals and metals occurring in the greenstones also appear in meteorites.

For a long time, these meteorites, which are considered the debris of other world bodies, have been a mystery for rock specialists. A person may consider that another world body must also contain granite, gneiss, schist, lime, and so on, and then ask: why do these rocks not *also* appear in meteorites?

Rudolf Steiner gave significant hints for the solution to this puzzling question in a lecture series given in 1923.[14] He explains there that the substance of meteorites originates from the *sun*. This substance is ejected by the sun from the area of so-called sunspots in such a fine state that one must speak of a *light-state* of matter. This fact, established by Rudolf Steiner's research, was confirmed twenty years later through spectroscopic investigations of the sunspots by the Scandinavian researchers Bengt Edlen and B. Strömgren using scientific methods. Therefore it may be considered certain that the materiality of meteorites originates from the sun and reaches the earth in definite ways and from definite directions. The author has reported in detail about this subject in Chapter 12 of his book "The Living Earth," which will not be further discussed here.

*From the described, it is clear that a comprehensive life process during the "Sun-stage" led to the formation of the "greenstones" and is still active today in metamorphosed form as a concrete process between sun and earth that leads to the formation of meteors and meteoric iron showers.* The same

---

[14] Steiner, R., *Supersensible Man*, Lecture III, Five Lectures given in The Hague, 13th to 18th November, 1923, Anthroposophical Publishing Company, London, 1961.

substance repeatedly appears in meteorites, namely that remarkable rock configuration we know as the ancient greenstones. Since the beginning of time, nothing appears other than what has to do with the sun. For this reason, granite, gneiss, schist, limestone, and other similar rocks never appear in the meteorites.

Through these meteor showers, sun substance is added almost without interruption to the earth and its living beings, but nevertheless in a living rhythm. The substance, which arrives in coarse pieces or as the finest cosmic dust, is predominantly iron, magnesium, and silica. It achieves the perfection of a gemstone in the olivine-chrysolite. All three substances are such that they unlock the earth's life materially, soulfully, and spiritually to the influence of the cosmos. Silica and magnesium are fundamental substances for the existence of plants because they convey sunlight into the plant. Iron and silica are the substances that make the human being's blood-perfused skin a sense organ. So in the formation of the eyes, we have that highest differentiation of the sense organs, which open the human soul and spirit to the world. The words of Goethe, "If the eye was not sun-like, it could never see the sun," expresses a profound truth. The eye is born entirely of the sun's activity. This is the same sun-activity to which the chrysolite owes its existence and ongoing development. Here lies the deeper reason why chrysolite-olivine has been associated with the eye since the earliest times. Rudolf Steiner confirmed this relationship with his research.

In its non-precious form, olivine is extremely common in greenstones and also forms separate rock varieties as peridotite. The most beautiful chrysolites known are also found in such peridotite. They are found on the small volcanic island St. John (Zabargad) off the Red Sea's Egyptian coast. Chrysolites were quarried there in pre-Christian times and used in the temples of Egypt and the Near East. However, their discovery was kept secret for centuries, and until 1900 the location from which the

Diamond Crystal
in Kimberlite
South Africa

Ruby Crystal in marble, Thailand
Star Ruby right,
top right cabochon cut

Sapphire Crystal, Sri Lanka
Star Sapphire right,
top left cabochon cut

Spinel Crystal in Marble, India

Figure 27: Artistic representations of gemstones discussed in this book.

Beryl Crystals, Brazil

Aquamarine Crystals in Pegmatite, Madagascar

Emerald Crystals in Mica Schist, Habachtal, Tauern, Austria

Topaz Crystals in Pegmatite, Brazil

Zircon Crystals in Marble, Hammond, New York

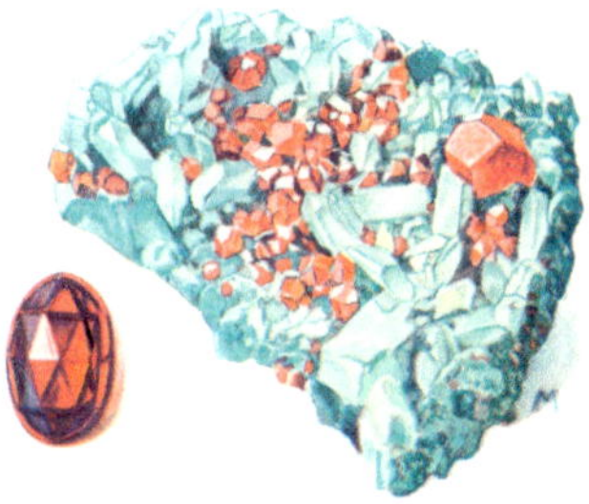

Garnet Crystals in Gneiss, Alps

Figure 28: Artistic representations of gemstones discussed in this book.

magnificent stones in the treasures of kings and churches came was unknown.

The island of St. John is basically the only place that provides really valuable stones. The color of the Egyptian stones is not surpassed by the few deposits elsewhere on earth. Therefore good chrysolites are by far rarer than good emeralds or even diamonds. They do not reach too high a price only because they are not resistant to wear. Polishable olivines are sometimes found in basalts from the Czech Republic (Bohemia) and those of the Eifel Mountains in the Rhineland Palatinate, west of Frankfurt, Germany. Dark bottle-green olivines come from British Columbia; very small ones are found in black lava sand in Hawaii. In a canyon in Arizona in 1904, some fine large olivine stones of up to 50 grams were discovered. In northwestern New Mexico, Arizona, and the Nile, olivine is also found in placer installations. Of these last three localities, only the American ones are significant.

The *garnet* accompanying the diamond, the so-called cape ruby or pyrope, is only one member of this large and multicolored gemstone family. Red garnets are commonly known. However, this gemstone also exists in violet-red (almandine), cinnamon-colored (hessonite), blood-red (pyrope, to which the cape ruby also belongs), yellow to reddish-brown (spessartine), and green (demantoid and andradite). Another green garnet is the grossular (gooseberry garnet). This variety of colors, as with all gemstones, is connected to the metals that give them color. Thus, we know about the *calcium-aluminum garnets*, which contain some iron and chromium and can be white, light green, cinnamon-colored, and hyacinth-red. To these belong the grossular and the hessonite. The *calcium-iron garnet* contains no aluminum, therefore more iron: it is colorless, often brown, green, or even black. To this group belong the sparkling demantoid and the topazolite.

Then there are the *magnesium-aluminum garnets*, which contain some iron and chrome. To these belong the bohemian pyrope, cape ruby. This group of garnets is best known for its blood-red, sometimes slightly brownish color.

The *iron-aluminum garnet* exhibits a bluish blood-red in its most precious form and is called almandine.

*Manganese aluminum garnets* always contain some iron; they are yellow or reddish-brown, orange-yellow, and are called spessartine when they are reddish-brown.

The *calcium-chromium garnet* or uvarovite is dark emerald green but forms such small crystals that it doesn't qualify as a gemstone. Only recently has it been found in coarse, fine-grained masses that are occasionally cut.

All garnets are silicates, like the beryls, topazes, and tourmalines. They are harder than rock crystals and, depending on the metal content, exceed a specific gravity of four, so they are quite heavy. Their material composition is almost as complicated and varied as that of the tourmalines. But their crystal form is characteristic and uniform for all colors: the rhombic dodecahedron. The crystals can be very small but can also reach and sometimes exceed the size of a walnut.

The garnet is widespread throughout the world as a rock-forming mineral. Finding garnet somewhere in the older rock types of a mountain range is not an adventure. There are countless places in the central Alps where garnet-bearing rock forms entire mountains. Its home is the green and gray schists, where it literally blossoms together with the bizarre forms of hornblende, green chrome mica, green chlorite, and blue cyanite. But it can also be embedded in schistose granulite (a white rock of fine-grained quartz and feldspar), marble and dolomite, and often in green serpentine.

A noticeable phenomenon of garnet is that it is mostly found in dense rock and only rarely forms in druses, fissures, and pockets. This is especially true for the precious red varieties. Its crystals, formed on all sides, sit like ripe fruits in the developing rock and clearly show that the surrounding rock mass hardened only *after* completely forming the garnet crystals. They are formed as crystals while "floating." It is, therefore, also difficult to extract such crystals from the rock. It is only possible to extract the garnet from soft chloritic schists, soft mica schists, and serpentine. Therefore, most garnet is found in weathering and disintegrated mother rocks and is obtained from placer installations. Due to its hardness and heaviness, this gemstone separates from the rock's lighter parts before forming true garnet sands.

It is always surprising to see the color harmony between the garnet and the rock in which it is incorporated. There exist in these places the most diverse rock types and mineral associations. The contrast between the red-brown-yellow tones of the garnets and the grayish emerald green of the mother rock is always striking. The polar opposite is the green garnets that appear mainly where the mother rock (serpentine) contains a lot of chromium, which gives the intense color. The most beautiful red-green polarity can be seen between the gemstone and its mother rock in the remarkable *eclogites*. This rock consists of densely matted crystals of leek to emerald green hornblende, into which smaller or larger blood-red garnet crystals are firmly embedded. This eclogite is also considered by petrology as one of our rock world's most enigmatic formations. It is tremendously dense and heavy due to its iron content, and at the same time, not at all weather-resistant. This condition results from the divalent iron in the green hornblende that is particularly sensitive to atmospheric oxygen. This stone immediately starts to rust if it is wet and exposed to the air. The eclogite is very close to the greenstones and is found in relatively few sites. In this type of rock, of which the garnet is so characteristic, one has a faithful image of the serene harmony in the mineral-plant life of the sun-stage of the earth. We feel the same harmony when looking at a densely green apple tree with ripe, red-cheeked fruit. But you can also think of a blooming bush of dark red roses.

It may seem absurd to use such images to "understand" garnet. However, it must be considered that the entire rock world is a real-world of pictures, which contains only the faint shadows of a world once filled with light and warmth. One can get to these shadows with chemical formulas, high pressures, and glowing temperatures to unravel them. But whosoever does this acts no differently from someone who cracks a nut with a thousand-ton press to determine how the nut developed.

This red-green polarity, where it appears in the individual garnet and in the garnet's relationship to the mother rock, can be looked at from another angle. The *red* and red-brown tones of most garnets come about through that form of iron the chemist calls *tri*valent iron. We have already learned about this form of iron with hematite. The other form of this metal, the so-called *di*valent iron, forms *green* compounds. Taking eclogite as an example, we can see that the iron is in the divalent (more plant-like) form in the *green* groundmass of the rock, while in the *red* garnets, the iron is trivalent and thus has the color of red blood. We have mentioned that the eclogite is not weather resistant, that it rusts. In other words, the divalent iron is oxidized by the oxygen and carbon dioxide content in the air, and with added moisture changes into the trivalent form. The very obvious conclusion is that the eclogite's green groundmass cannot have originated in today's ordinary air. The atmosphere must have been practically free of oxygen during the formation of these green hornblende minerals; otherwise, all the iron would have been oxidized. Nevertheless, trivalent iron is found in the garnet crystals embedded in the eclogite.

This coexistence of the two forms of iron is widespread and found in other mineral associations. Here, just as in life processes, they point to a higher-level principle where chemical laws of matter dominate.

However, this observation implies that the divalent and trivalent iron is used as a tool by a living organism and can *coexist* within a life process independently of the inorganic laws of matter. This higher-level principle is nothing other than the "life" of the ancient mineral-plants, which we have already discussed. In the sphere of their "life-activity," such "unstable" mineral formations could build up organically, and when exposed to the effects of the atmosphere – the water and the air – would decay and "decompose" just like a corpse.

The red garnet belongs to those remarkable stones called carbuncles in ancient and medieval times. By this name, a certain blood-like color was meant that is independent of the rock's substance. In such stones, one saw the purified, chaste, and passionless counterpart of human blood. Whoever wore such a stone could always visualize that it was a goal to transform their own blood in such a way that it would develop the same pure and desireless power as the clear, transparent gemstone.

The justification of such an idea can only be recognized from Rudolf Steiner's research which has shown that the earth's siliceous forms emerged from a pure, light-woven mineral-plant "life." From this life, the gemstones unfolded like flowers and fruits.

The places where garnets are found are widely spread over the earth. Of the numerous places where this gemstone is washed out of the rubble by placer installations or broken out of the loose rock, only those of significance are mentioned here. *Almandine*, which takes its name from its oldest site, Alabanda, in Turkey, comes today mainly from the gemstone placers of Sri Lanka. In Near East India, Rajputana, Sarwa, and other places in Rajasthan, colored stones are found in weathered gneiss and mica-schist. In Afghanistan, South Africa, Brazil, and Madagascar, almandine is also found by placer installations. Alaska and Greenland should also be mentioned. In the past, crystals up to five centimeters in size were extracted from the mica and chlorite schists of the Ziller Valley in Tyrol.

The pyrope or magnesium-aluminum garnet has, since the oldest times, appeared most beautifully in the low mountain range of the Bohemian Forest in the Czech Republic. It is blood red and is rarely found more than eight millimeters in diameter. There were times when large garnets from Bohemia fetched the price of rubies. After discovering the South African diamond deposits with their large pyrope garnets (cape ruby), Bohemian Forest production quickly declined, especially as this gemstone became increasingly out of fashion. After these two localities, which are still the most important in terms of quality, there are Sri Lanka, East India, Australia, and several places in the southern USA. Between the two world wars, Madagascar provided a beautiful dark rose-red garnet that is brighter than other pyropes, but it has now disappeared from the trade. Garnets from North Kordofan, Sudan, often exhibit a pure crimson color and have especially large crystals.

*Hessonite* (cinnamon garnet) is found in Sri Lanka's gemstone placers, in East Africa, the USA (California and Virginia), and the Gotthard massif in Switzerland. The cinnamon to hyacinth hues of this gemstone are highly prized. The greenish varieties of this calcium-aluminum-silicate garnet group, the so-called grossular (from grossularia, the gooseberry), come from Rajasthan in India, Madagascar, South Africa, and in North America from Canada, California, and Maine. Other localities are in Europe at the Băiţa mining district, Romania, and Monte Rosa in the Alps. Colored, opaque grossularite is ground into dome shapes, is often jade-like, and comes from Siberia's Vilyui River region.

*Demantoid* (andradite) is a calcium-iron garnet found exclusively in the Urals. It has a special bright green color and is characterized by its high light refraction. Its hardness is lower than that of the other garnets. It is the rarest and most valuable garnet. In Transvaal in South Africa, a coarse, minutely crystalline demantoid of good color is found and cut as Transvaal jade or Transvaal nephrite.

The manganese-aluminum garnet *spessartine*, found in crystals up to seven centimeters in size in tourmaline pegmatites in Southern California, is the only garnet that occurs outside of the greenstones. The deep red to brown-red crystals of this gemstone are rarely transparent, as their coloration is very dense. It is also found orange-yellow in Madagascar and, in Brazil where it occurs massively as spessartine rock. The weathered material can be extracted as manganese ore. This gemstone gets its name from its oldest site near Aschaffenburg in the Spessart Mountains. The author found a very beautiful spessartine gemstone in 1931 at the Große Mösele, a mountain in the Zillertal Alps. The dark red, beautifully formed crystals were embedded in a green hornblende rock together with snow-white calcite.

## 16. Jet and Amber

In the preceding chapters, the attempt was made to describe the formation processes of the gemstones as characteristic of states of the earth and its realms of nature in the past. Still, it is extraordinarily difficult to develop correct ideas about these states and forms of life from the past because today, they are no longer visible. An observer of the mineral world is in the same position as the anatomist facing a corpse to be dissected. The observer can exactly determine and describe the place, form, and structure of the organs, but it will hardly be possible for the observer to form correct ideas about the origin and function of these organs in life. This is because the primary element of a formation becoming visible – an organ, an organism, a plant, or a mineral – is always an *activity* that takes place completely in the realm of pure *forces*. It does not matter how these forces are initially described. All phenomena, be they organic or inorganic, the growth of a plant or development of a crystal, occur in an invisible but organized field with a certain configuration of forces.

The objection that one cannot compare a corpse with a crystal is increasingly invalid since it can be clearly shown that mineral matter is subject to decay. Rudolf Steiner has often called the mineral the skeletal system of the living earth. In addition, the atmosphere, water, warmth, light, plant, animal, and human beings all belong to this *living* earth. The mineral realm, the orderly and intricate skeletal system of the once much more living earth, is just as much the remnant of organic life as the human or animal skeleton. As we know, the skeleton can last a very long time because it is the most mineral part of the human and animal organization.

Suppose someone finds a bone or an entire skeleton somewhere; they won't claim that this formation has a purely mineral nature, although it consists entirely of calcium phosphate. By its *structure,* the observer recognizes with certainty its organic origin. But if the observer examines this structure with the microscope, they would discover forms in the mineral world and find crystals of calcium phosphate that look exactly like the crystals of the mineral calcium phosphate (apatite) in the earth's primary rocks. The mineralogist who knocks around on rock and stone in a quarry, mine, or on the top of a mountain, and draws quick conclusions about the origin of minerals, is exactly like the metaphorical researcher who concludes from the crystals in a fossil bone that it is only a mineral structure.

The mineral realm, the skeleton of the once much more living earth, is a gigantic structure that points to equally gigantic life processes. Whoever does not attempt to grasp this giganticness in the view of the rock world will not go beyond the "microscopic." Thus, it is not without reason that Rudolf Steiner repeatedly urges that when the researcher contemplates the smallest single parts, they must penetrate deeper into the *larger context*. Otherwise, the connection with reality is lost.

The life of the earth has gradually developed from gigantic, comprehensive states of life, which were not differentiated into individual forms, to the present differentiation of the nature realms and their forms. Those world forces active in the old states of life have transformed in connection with the development of individual forms and are active today on completely different levels of existence.

Therefore new formation of rocks and minerals slowly comes to an end, while at the same time, the disintegration, weathering, and decay of the mineral realm begin.

Concurrently with the beginning of this period of decay, the formation of two more or less precious deposits from the already existing plant realm occur; *jet* (gagates or black amber) and the amber itself.

The formation of these two substances is widely separated in time. The jet is found in the Posidonia shale of the early (black) Jurassic, which also holds the famous ichthyosaurs at the foot of the Swabian Alps in the middle (brown) Jurassic and the Cretaceous layers. These strata belong to the great final chapter in the history of rocks, which comprises mainly enormous calcareous deposits from animal life.

Jet is nothing other than very dense, hard coal of deep velvet-black color. Its fracture is conchoidal and shiny. In the past, polished material was used for mourning jewelry and rosaries; it can be nicely polished and then has a warm waxy luster.

The origin of jet is shrouded in some uncertainty. Since one sometimes finds pieces that still show a faint woody structure, it is assumed that wood was embedded in rotting mud and therefore strongly bituminized. In any case, the original organic substance has undergone an extraordinary shrinking process. Along with the jet, pieces of fossilized wood are found that have been completely transformed into sulfurous-iron ore (pyrite or marcasite). This finding indicates that, indeed, jet can originate from wood. However, one should not imagine that this wood looked exactly like today's wood. It is precisely the phenomena of strong shrinkage, mineralization, and silicification that point to woods of great porosity and softness. The flora of *this* period was not yet developed to the density of present-day forms.

There are many places where jet is found in abundance. They all played a greater role in the past than in the present, when black synthetic resins "replace" the jet. Since the material can be carved, ground, and turned, small industries were located near these finding places until the First World War. These were in Yorkshire, England, southern France, Spain, and at the foot of the Swabian Alps in Germany. The jewelry industry of Schwäbisch Gmünd in Germany started in the past with the processing of jet. In Utah in North America, there are such huge jet deposits that there is enough to supply the whole world for centuries.

*Amber* is younger than jet. We know with certainty that this fossil resin originates from large coniferous forests present in the early Tertiary, mainly in the Baltic Sea area and some other places. From the immense quantity of amber found in the Baltic Sea area and at the German North Sea coast (between 1900 and 1930 some 10,000 tons or 10 million kilograms), it is possible to conclude that not only were there very large forest areas but also that the trees were of a special vitality that excreted this resin.

Figure 29: Baltic Sea Amber in a variety of colors.
(Photo by M. Porsina, Courtesy of Wikimedia Commons)[xxi]

In some respects, it is quite mysterious how these quantities of resin came about. We know from our present-day conifers that, on the whole, they only excrete resin when they are injured. The tree cannot long withstand a major injury before it literally "bleeds out" and dies. The resins that get into the soil of our forests in this way nowadays weather and crumble quickly and do not show the slightest tendency of becoming "fossilized." Therefore, it is difficult to imagine that the amber originated from a "typical" coniferous tree resin. There are numerous inclusions of insects of different kinds, which show clearly that the amber was soft and thick immediately after its deposition. Also, plant inclusions such as wood dust, pieces of wood, and conifer needles point to it. The size of the resin pieces is also astonishing. The largest piece of amber ever found weighed 18 pounds!

On the other hand, what became of the massive amounts of wood from those coniferous trees? The amber lies in *deeper* layers than the tertiary brown coal, and in this brown coal, amber-like resins are found only as a great rarity.

Taking all of this in, one gets the impression that the amber is the product of a natural catastrophe that happened over a short period. The trees that excreted those immense quantities of resin were probably severely damaged in their crowns by storms causing the resin to flow out in large quantities. Immediately after exudation, the resin must have quickly been excluded from the air; otherwise, it would have dried up and decayed. The storage of the amber between clayey sands and clayey layers indicates that it was washed away and deposited in a completely different place than the wood it came from, which was eventually subject to complete decomposition.

There is no doubt that the above-mentioned natural disaster is a chapter of the great Atlantean catastrophe, which destroyed the continent between North America and Europe over a longer time. Today's geological conditions in the Baltic Sea region point to large changes during this time.

The colors and structures of the amber confirm the above-described occurrences. Only a small part of the amber is completely clear because it was underwater immediately after flowing from its source. Most amber is turbid because it has undergone a drying out process over an undetermined period. By this drying out, one must understand the evaporation of essential oils, similar to turpentine. As a result, the amber becomes turbid, opaque, or even white. The white amber is completely foamy and is called bone amber. The white color comes from fine bubbles between one-hundredth and one-ten-thousandths of a millimeter in diameter. These fine bubbles produce the well-known wax and honey-yellow and red-brown coloring. They also produce sky-blue and green, but these colors are precious rarities.

Another resin is found together with the amber, which differs from the latter in that it contains *no* sulfur like common amber. This resin is called gedanite or brittle amber. Gedanite is very similar to amber but is more soluble in hot linseed oil and other solvents. Its value is lower because of its brittleness.

Other amber sites in the same Tertiary strata are Romania, Sicily, Myanmar, the Dominican Republic, and Mexico. These amber types differ from the East Prussian amber, mainly because almost all of them have a strong petroleum-like fluorescence. The appearance of a blue-green fluorescence is a great rarity in East Prussian amber and increases its value considerably.

Where found, the amber is dug in open pits and extracted by sluicing. The amber-bearing strata lie about 30 to 40 meters below the surface. On the Baltic Sea coast, especially on the west coast of the Sambia Peninsula (near Kaliningrad, Russia), amber is very abundant, although amber is also "fished" or collected by amber hunters on the mudflats of the North Sea coast after stormy seas. This "sea amber" is the most valuable and durable because it doesn't have a crust from weathering.

Amber has been used by human beings since the Stone Age. In the Orient, it originally served only cultic purposes. Nero had an amber expedition undertaken to gain possession of the precious material. In Germany, from the 16th to the 18th century, hundreds of articles for daily use were made of amber. Even boxes, flasks, and ornamental vessels were made entirely of amber.

The most outstanding property of this fossil resin is that it is negatively charged by rubbing. From the Greek name "electron" for amber comes the term "electricity."

Another property, based on widespread folk experience, is the analgesic effect of amber for teething children. This effect is probably due to the release of minute quantities of odorous substances formed when the amber is worn on warm skin. The medical advice of Rudolf Steiner goes in the same direction: he recommended burning incense with amber to address phantom pains of amputated limbs.

## 17. Pearls and Corals

We leave the past with the jet and amber and come into the present with *pearls* and *corals*.

It seems significant that today the formation of precious substances from organic matter only occurs in the calcium metabolism of lower animals. This is so because, as a *substance,* calcium is connected with life to a much higher degree as a scaffolding substance or shell than we could say about any other substance.

Mussels and coral polyps, the two living organisms that yield the pearl and the coral tree, are known from the earth's older strata. But these two animal genera have very different ways of dealing with calcium. While the pearl mussel grows underwater attached to the bottom and forms its shells *from inside to outside,* the coral, especially the precious coral, builds its trees freely into the surrounding water perpendicular to and away from its attachment point. From this, it is clear that for the colony of coral polyps building their little trees by the thousands, the earth with its center of gravity in the middle does not exist. In addition, the coral processes the calcium *from outside to inside,* in contrast to the pearl mussel.

If one wants to compare these two processes with the calcium process taking place in the growing human being, then one would have to point to the formation of the skull (the mussel process from inside to outside) and the formation of the limb bones (the coral process from outside to inside).

But we can also study this contrast of calcium deposition in the earth as a whole. We see the principle that forms the mussel shell realized on a large scale in the shell limestone formations, which will be discussed later. The structure of the shell limestone formation is essentially rhythmic, alternating limestone beds with dark, schistose marl layers containing a carbonaceous, coal-like substance.

Figure 30: Soft Red Coral with polyps.
(Photo by Nhobgood Nick Hobgood, Courtesy of Wikimedia Commons)[xxii]

In the Jurassic Formation, we have a completely different structure of limestone layers. They are mighty, thick-banked reefs standing in the landscape as mesas and luminous rock faces. The outermost pinnacles of these reefs are often formed entirely by fossil corals.

If one studies the finer structure of the mussel shell, three prominent material or form elements can be observed. First, the mussel shell and the pearl have calcium carbonate as a basic mineral substance present at 92 percent on average. Second, this calcium carbonate occurs in two forms: in prisms about one-thousandth of a millimeter thick and in platelets about several ten-thousandths of a millimeter thick. Third, these platelets form the beautiful mother-of-pearl layer that is well known from other non-pearl-forming shells.

The prisms standing perpendicular to the shell surface or pearl surface are different in their mineralogical crystal structure in the shells of the sea pearl oyster and the freshwater river pearl mussel. The sea pearl oyster forms *trigonal calcite* in its prisms, whereas the river pearl mussel forms *rhombic calcite or aragonite*. Mother-of-pearl layers are formed of aragonite in both species of mollusk. Between the prisms and the layers, very fine films of an organic substance called conchiolin are deposited. The conchiolin is a so-called scaffold protein and has a close material relationship with the chitin in insect bodies.

Here we see how a living organism confidently masters the various mineral forms of calcium. We know the trigonal and rhombic (*aragonite*) calcite above all from the mineral world. These two forms of calcium develop in inorganic nature from the original calcium gel when certain environmental conditions of warmth are present. Calcium is the "mineral" substance having the most crystallization *habits* to exhibit, and we know more than 800 of them.

The mussel has completely different cells in its "mantel," close to its shell—those that only produce conchiolin, those that form calcite prisms, and those producing mother-of-pearl layers. Everything happens with an unbelievable exactness, an exact order, and, yes, one can only say, incomprehensible wisdom. This insight is the deeper reason why Rudolf Steiner presents the *oyster* in his medical lectures as something very special. This common oyster does not form pearls, but it is closely related to the pearl mussel. Rudolf Steiner says of the oyster that it can only live by pushing the calcium outward. It has bonded itself so strongly with the earth that it must constantly repel the earth from itself through its shell formation. Thus it develops an extraordinary inward vitality and reproductive power. River pearl mussels can live for 100 years, sea pearl oysters only for approximately ten to twenty years because they live in warmer water.

Figure 31: Mother of Pearl with accrued pearls.
(Photo by Hannes Grobe, Courtesy of Wikimedia Commons)[xxiii]

Rudolf Steiner connects this process of calcium rejection with *human thinking*. The thinking ability comes about because human beings *do not* let formative forces flow into the material of their bodies and thus do not continue to grow for the rest of their lives like an oyster; instead, they hold these forces back and let them shine as inner thoughts. The oysters, however, or the mussels in general, seal their "thoughts" into the calcium. If they could hold back what they put into the design of their shells, they would be very clever animals. This cleverness, this wisdom appears in the wonderful structure of the shells. It is really "world thoughts" that are revealed through their activity. And whoever wears a pearl must actually say that this dull shimmering sphere is a picture of the world's thoughts that work all around the earth and have become apparent through the pearl mussel.

The pearl is formed by a foreign body getting between the shell and the animal's mantle, and the animal is thereby induced to cover the foreign body with shell substance. Depending on the position of the foreign body on the mantle of the animal with its different cells, dark conchiolin pearls, light prismatic pearls, or the delicate mother-of-pearl pearls are formed. Besides the perfect sphere, which is the most precious type of pearl, pearls can be of any shape and are then called baroque pearls. Drop-shaped and oval forms are also popular. The most important thing is the "luster" of the pearl, the sheen, and the color, which are entirely dependent on the pearl's internal structure and its way of refracting and reflecting the light.

The home of the sea pearl oyster is in warmer seas. The most beautiful "Oriental pearls" come from the Persian Gulf and Sri Lanka. They are also found in the Red Sea on the African coast and the coasts along the Indian Ocean. Pearl mussel beds are also found on the north coast of Australia and the coasts of Central America. Insignificant deposits are located in the Indian Archipelago and near some islands of the South Seas. In the Sea of Japan, there are mussel beds near Sakhalin. South Japan has had a large cultured pearl industry for decades under the famous Japanese Mikimoto Kōkichi. His research has given us the most accurate knowledge about the life of pearl mussels and the formation of pearls.

The *river* pearl mussel (freshwater pearl) is another species found only in clear streams and rivers. Interestingly, it prefers *low*-calcium mountain waters and has been harvested in the creeks of the Odenwald, the Fichtel Mountains, the Bavarian

Forest, the Ore Mountains, and Upper Austria. River pearl mussels are also known in England, Scotland, Ireland, Scandinavia, and northern Russia. The existence of the river pearl mussel depends on the presence of certain fish, which harbor the juvenile larval stage of the mollusk in their gills. Due to the pollution of our waters, the river pearl mussel has become a rarity. The height of its harvesting belongs to the past. Today, the only significant harvesting of river pearls is in North America, where many rivers are home to pearl mussels.

The colors of pearls range from metallic black through the entire rainbow to purest white. Even though it is made of calcium, the pearl's hardness is higher than that of the ordinary mineral calcium. The concentric shell structure, interwoven with a radially-fibrous structure, causes this hardness and gives the pearl great mechanical strength.

The size of the pearls is extremely variable and is related to the size of the shell that produces them. Since sea pearl oysters live only about 12 years and river pearl mussels up to a hundred years, large freshwater pearls are more often found. But sea pearls can also reach weights of about 40 grams.

The *coral* stands in contrast to the mussel and the pearl in its formation, as already mentioned above. Although its life sphere is outside, it is indeed attached somewhere, and it works the calcium *inwardly,* forming a framework, an *evolutionary inner skeleton*. Something appears here in a lower animal that finds its perfection only in the higher animal and the human being.

It is important to note that the *forms* of calcium produced by the hundreds of coral species consistently have a *plant-like* character. There are cup corals, organ corals, fan corals, and others. The precious coral, which occurs in red, white, and mottled red and white, forms little trees branched spatially and reaches an average size of 20 to 40 centimeters with branches 2 to 4 centimeters thick. Larger coral canes have been found in the Sea of Japan, up to 1 meter in height and 20 to 30 kilograms in weight.

In their burrows, the corals deposit enormous amounts of calcium carbonate on the seabed. When the seabed sinks, they continue to build upward, creating entire coral reefs that can be dangerous to ships because of their solidity. When the seabed rises again in geological time, the reef comes to the water's surface and is covered with plant debris by the current. This is the process of how the coral islands and atolls of the South Seas are formed, and they are often only destroyed by the most violent storms. From these phenomena, it can be seen that corals are ancient earth-builders. Numerous coral species, which inhabited warm seas on a much larger scale, created enormous land areas in past times. In the White (Upper) Jura formation (southern Germany), you can partially see such land areas.

The precious coral does not belong to those species that have formed land areas in the past. Its occurrence is limited only to relatively narrowly defined marine areas: the Mediterranean Sea, Bay of Biscay, Canary Islands, Irish coast (Gulf Stream), Mauritius, the Malay Archipelago, and Japan. In water depths from 3 to 300 meters, the red and occasionally lighter colored corals form fields and banks. The living trees are covered with a soft bark that is easily scraped off the harvested coral. The coral is then revealed, which is finely grooved on the outside and has wart-like dimples in which the polyp animals resided. The coral tree consists of about 83 to 86 percent calcium carbonate, deposited as calcite in fine crystalline form. In cross-section, it is possible to see under the microscope that the structure is concentrically ringed like a tree and radially, corresponding to a wood's medullary rays. Besides calcium, the coral contains magnesium, silica, iodine, and about 4 percent iron oxide. Here, the beautiful shades of red come from carotenoid pigments and not from iron. The coral also contains an organic putty substance similar to the pearl but of a different type, resembling horn. This putty content, like the water content, is much less than that of the pearl. The coral as a whole is more *mineral* than the pearl. The colors range from pure white to dark red. Japanese moro corals from Ryukyu Island are particularly dark. Other Japanese species have a white core and are red on the outside, and vice versa. In the Mediterranean, black corals with a red core and also spotted and striped corals are known.

As we mentioned above, the coral has an evolutionary relationship with the internal skeleton. However, the very nature of coral will be understood when considering that its shape imitates the form of blood vessels. This expresses that the calcium in the higher animal solidified in bone is already to a certain extent solidified in the vascular system. Thus, the coral represents the calcium that is still contained in blood and is on its way to become bone. From such a point of view, it is understandable that the *red* coral is a substance that acts on the blood and the blood vessels. In this sense, Rudolf Steiner also indicated it for use in certain remedies.

## 18. A Few Words about "Synthetic" Gemstones

The art of fabricating transparent substances such as glass fluxes was known early on by the ancient Egyptians. Sand, natural soda, and shells were melted into easy-flowing mixtures and colored by adding metal ashes. Originally, this art was a temple secret, just as utensils and jewelry made of such glass exclusively served cultic purposes.

Up to the time of the medieval alchemists, when magnificent church windows with stained glass had long been created, the art of making "vitrum" (Latin, glass) was considered the pinnacle for working with the material.

This glass has survived almost unchanged regarding its manufacturing processes and composition; it is essentially a new human creation with no counterpart in nature. Certainly, volcanic "glasses" exist, such as the obsidians occasionally made into jewelry, but these "natural glasses" are rich in alumina and silica and very difficult to melt.

Although humanity has known about glass fabrication for perhaps three millennia, no one thought of making artificial and colored glass a "substitute" for real gemstones until the 19th century A.D. Therefore, the types of "synthetic gemstones" fabricated today in unlimited quantities for cheap and kitschy mass-marketed jewelry will not be discussed here.

It is only significant that at about the same time as people began to imitate gemstones in the manner described, the first attempts were also made to produce "real gemstones" artificially. They tried to produce by technical means crystals that were chemically like natural crystals.

Given the high value of precious corundum (ruby and sapphire), the first efforts were directed almost exclusively to the artificial fabrication of the ruby. Detailed chemical studies of gemstones, which had already been carried out in the first half of the 19th century, made it possible to use a broad range of melting and conversion processes to achieve the goal. All of these procedures were successful. The only difference was that the effort was greater than the gain in most of them. Of all these attempts, a single process developed by the Frenchman Verneuil prevailed. It uses an ingenious device to introduce the purest alumina as a powder with about 2.5 percent chromic acid into an oxyhydrogen flame. In the flame, which has a temperature of over 2000 degrees, the alumina melts into tiny droplets, which gradually collect on a refractory pin. After several hours, it has melted together to form a "pear" about 6 centimeters long and 15 to 20 millimeters thick. This "pear" has a thoroughly crystalline structure, is flawlessly clear, and is of a pure ruby red color when the right powder composition is used.

Suppose such a "pear" of artificial ruby is cut. In that case, the first thing that becomes apparent is the hardness of the "artificial" material, which is much *harder* than the natural version, for all its superficial similarity to the natural stone.

*Under the microscope and with further optical examination methods, the expert can easily determine that the artificial gemstone has a different internal structure than the natural stone despite color similarity and high hardness.*

Following a technically improved Verneuil method, these synthetic gemstones are manufactured today in some countries. It is possible to produce completely colorless gems using the purest aluminum ore. Any color can be produced by adding the most diverse metals. Even the mysterious red-green color change of alexandrite has been produced by adding vanadium to the molten alumina. More recently, "synthetic star sapphires" have been produced.

The same Verneuil process also produces synthetic *spinel*. Natural spinel consists of a "molecular" ratio of approximately equal parts alumina and magnesium oxide. The synthetic spinel contains much *more* alumina (aluminum oxide) since masses with the molecular ratio of natural spinel cannot be melted into perfect pears. Materially, therefore, artificial spinel is, strictly speaking, something quite different from the natural stone.

These artificial spinels also have a high hardness and can be colored as desired. Therefore, they are used more for jewelry than as artificial corundum.

At this point, to the jewelers' credit, it must be said that artificial gemstones are only traded as *such*. The greatest part of artificial gemstones, especially corundum, is used for technical purposes such as bearings in clocks and other precision instruments.

In the pursuit of an artificial gemstone that surpasses the diamond's fire, pure titanium oxide was melted in the Verneuil furnace. Titanium, which we have already learned about connected with the coloring of various gemstones, is a metal related to iron, yet light and very hard. Recently, it has begun to play an increasingly important role in technology. This metal oxide occurs in nature as rutile, anatase, and brookite, minerals found in granite in fissures with rock crystal and other minerals. Natural titanium oxide, called "rutile," very rarely has a gemstone character.

A new synthetic gemstone can be made from an almost colorless rutile that has been artificially produced. When cut in the brilliant form, it has much greater light refraction and about seven times greater color dispersion than the diamond cut into the same shape. This "super diamond," as it could be called, has only one big disadvantage, it is softer than rock crystal and is about the hardness of feldspar. As a result, it wears out easily and loses much of its fire. To compensate for this deficiency, the finished cut stone is coated with a harder material, such as quartz or corundum, in a high vacuum by vapor deposition. Significantly, this synthetic rutile develops the full intensity of its color dispersion only at night in artificial light. In America, it was given the name "Titania, the night stone."

Another synthetic gemstone produced in America by the Verneuil process is the substance strontium titanate. It is a compound of titanium oxide with strontium oxide. Strontium is an alkaline earth metal closely related to the calcium of ordinary lime. As sky-blue celestine, the sulfuric acid salt of strontium is found in some lime formations and as needle-like (acicular) coverings in the Girgenti sulfur mines of Sicily.

The "Starilian," a commercial term for synthetic strontium titanate, has no counterpart in nature; thus, it is a genuine new creation of the human being. The strontium titanate is absolutely colorless in contrast to synthetic rutile and has almost the same hardness as the latter. Its brilliance, fire, and transparency are similar to a diamond's since the color dispersion is not as intense as synthetic rutile. This new "gemstone," Starilian, is not sold as a rough stone in the form of fused pears, nor as a cut, loose stone. Rather, it is set in platinum jewelry by the fabrication company and sold only in this way.

For some time, a German company has been selling an artificial Lapis Lazuli, which has only the color in common with the natural stone. The stone is made of sintered corundum or spinel and is colored by suitable metal additives or artificial ultramarine blue. The color is an excellent lapis blue, even with the characteristic star-like inclusions of pyrite (iron sulfide). Furthermore, the hardness is much higher than that of the natural stone since it is corundum or spinel.

The synthetic gemstones described in the preceding paragraphs can be considered close relatives of glass in their technical fabrication because they emerge from a melting furnace. However, there are now also fabrication processes that result in real crystals, which can be equated with natural crystals. The only fine distinction that must be made here is that the melted pears coming from the Verneuil furnace are *internally* crystalline, but externally they do not have such exact boundaries as the natural crystals. Externally, such pears may look like glass, although they differ by their *inner* crystalline structure. Real glass is *not* crystalline but a solidified *gel*, a colloid. This condition can be seen because large and heavy window panes become thicker and thicker at the bottom over the years. After all, real glass still "flows."

The artificial fabrication of real gemstone *crystals* without high *melting* temperatures goes back to extensive experiments in the $19^{th}$ century. In the years before World War II, a German company succeeded in producing emerald *crystals* from *aqueous* solutions of the emerald components.

The procedure is to heat the individual components of emerald known from analyses – silica, beryllium, alumina, and chromium oxide – together with pure water in a pressure-resistant vessel to about 400 to 500 degrees. The superheated water with a pressure of up to 1,000 atmospheres "dissolves" the substances that are not readily soluble in water and form colloids. If this colloidal state of solution is maintained for weeks and months, with close attention to pressure and temperature, true emerald crystals crystallize and settle in the upper, cooler part of the pressure vessel. However, in this process, it is usually necessary to add a so-called seed crystal to the solution to start the crystallizing process.

Using this very interesting technique, several thousand stones of up to three carats individual weight were produced monthly in San Francisco by an American chemist.

The same process is used by an important American company in the electrical industry to manufacture large, clear, and doubly terminated *rock crystals* used only for technical purposes as so-called control crystals for transmitters, ultrasonic generators, etc.

It is possible to produce quartz crystals up to 500 grams in weight *in a period of two to three months* using this hydrothermal technique. Although these artificial crystals are very expensive due to the enormous expenditure of time and energy, they pay off:

good emerald stones are rare, and there is a lack of perfect natural rock crystals suitable for technical purposes.

Apart from the fabrication method, things are basically very simple regarding the "valuation" of such artificial gemstones.

It should be noted that such artificial products appearing in many other fields are true offspring of our time and of our way of thinking. The human being has learned to handle substances through an intense occupation with sense-perceptible matter and the almost constant search for its properties. Ultimately, the gaze is directed to these properties, to the "what" of any substance. If, for example, we succeed in the fabrication of a material that can no longer burn and is also unbreakable, then something significant has certainly been achieved. But the layman has no idea what technical procedures bordering on magic would have to be used to produce the substance. The chemist, the engineer, knows about these procedures, but the consumer, who uses the artificial substance in any way, has no awareness of the "how" of the ways that led to it. There are hardly any more questions about *"how"* a substance comes about. And yet, the *ways* that lead to the formation of a substance are decisive for its *inner* qualities.

The decisive thing with all artificial products, be it gemstones or any other synthetic material with "supernatural" properties, is that a completely new world is created, which would not exist without human beings. Human beings must awaken to their responsibility regarding this just as towards the other natural realms. But since they have neither one nor the other, they are merely proud of their creation and speak of having outdone and "improved" nature. But what if the question would be raised whether the "nature" had ever had the intention to go beyond what it has "created?"

One thing is clear: nature has gone about creating her substances in a *totally different way* from today's human technology. The results of *her* way are the natural substances. The human being has been interested until today far more in the *result* of this "natural way" than in the way itself. The human being has only studied the properties of the result, namely the *properties* of matter, thereby making the great mistake of concluding from the properties how they came about.

When technology succeeds in producing an artificial gemstone by melting natural substances or letting crystals develop by heating a solution under enormous pressure for months, *the conclusion is that nature worked the same way*. In reality, the artificial production of crystals is *nothing more than an understanding of the need for great heat or high pressure to create a substance that is similar to a natural substance*.

In an unbiased study of natural substances, which proceeds with the utmost exactness, it must, however, be noted again and again that nature has gone *other ways*. Everything points to the fact that even the "dead" substances of our rock world had different chemical and physical properties in the past than they do in the present. This point is just as certain as the fact that there were other plants and other animals in former times, which the present ones have replaced. Thus, in reality, nature is in flux everywhere, even today, only with the difference that the human being has been involved with this flow for some time.

Paracelsus spoke of the fact that the physician and natural scientist must "go through the exam of nature." This means that they must learn how nature works. On the other hand, Rudolf Steiner says the human being can learn to work with the *ongoing flow of nature* and must go beyond *existing* nature.

Today, however, our entire ability lies in the realm of existing nature and its *existing* substances. We work only with what has become, without asking what *nature's* way is for the future. This way reveals itself only if one asks about the ways of nature in the *past* and *how* a substance came about.

From the question about "how" a substance came into being in the past, answers can arise about "how" to proceed toward the future.

Today's work with existing nature will lead to its destruction and the destruction of life in general if one cannot find ways to counterbalance and work with the ongoing flow of nature.

A person looking at a synthetic gemstone in its perfection can say: this is a product of human intelligence; the human being has produced this beautiful structure knowing *existing nature* very precisely. *It is a new thing in the world*.

But suppose this person holds a natural stone of the same extraordinary beauty in their other hand. In that case, it can be said it is a form of *emerging* nature, whose ways I can only guess, whose final secrets are still completely closed to me. *So this natural stone in the world is an ancient witness for everything, that in the end, has emerged from light itself.*

## Walther Cloos – Biography[15]

Walther Cloos was born in Darmstadt, Germany, on June 22, 1900. He was a husband, father, pharmacist, alchemist, Anthroposophist, lecturer, researcher, inventor, and author during his life. He died in Ulm on June 30, 1985.

He grew up in Darmstadt and studied pharmacy at the Stuttgart Technical University, attending lectures on mineralogy and geology and doing practical pharmacy training in the Spessart region and the southern Black Forest.

Walther Cloos was introduced to Rudolf Steiner and Anthroposophy through Joanna Thylmann's reading group after hearing Steiner speak at the independent academic course held in Stuttgart in March 1921. Later he would say that it was through the study of Steiner's written works that he gained a great appreciation for what he brought into the world.

After having completed his apprenticeship in pharmacy, he went to work in a chemical factory in Darmstadt. Shortly after Rudolf Steiner's death, in April 1925, Cloos joined the Institute of Clinical Medicine's laboratory staff in Stuttgart, the nucleus for what later would be the Weleda Company in Germany. There he started to develop the method for metal distillation which Rudolf Steiner had suggested would increase the medical actions of metals. These metallic mirror preparations, called 'metallicum preparatum,' have come to play an important role in the

---

[15] Adapted from the "Pioneers in Anthroposophic Pharmacy" PDF online at https://iaap-pharma.org/publications.

medicinal use of metals. Cloos was a true "Son of Cain." His task was to take earth's substance and transform it, developing "evolving nature" further for healing purposes. He would not shy away from extreme industrial procedures involving enormously high temperatures and high vacuum to cope with something as tremendous as the distillation of metals. He was also interested in medicinal plants grown by the Weleda company in Schwaebisch Gmuend, outside of Stuttgart. With the botanist Gerbert Grohmann and many farmers, he studied the ideal conditions and best habitats for different medicinal plants and their biodynamic cultivation. The main focus of his work at Weleda would be the manufacture of medicines made from minerals and metals. Cloos worked out some 200 procedures for these medicines.

In the 1960s and 1970s, Walther was part of a collaboration at Schwabisch Gmuend with Willhelm Pelikan, Oskar Schmiedel, Hans Krüeger and Theodor Schwenk. This would be the foundation for Weleda medicines later on. Cloos also kept in close contact with the medical profession, which provided opportunities for further collaboration.

## Also by Walther Cloos:

*The Living Origin of Rocks and Minerals*, published by Floris Books, 1977 (also published as *The Living Earth: The Organic Origin of Rocks and Minerals*, Lanthorn Press, 1970). Originally published in German as *Lebensstufen der Erde* by Verlag Freies Geistesleben, Stutgart, 1952.

## Endnotes and Photo Attributions

---

[i] Cover Image: Quartz, Tibet. Photo by JJ Harrison (https://www.jjharrison.com.au/) (https://commons.wikimedia.org/wiki/File:Quartz,_Tibet.jpg), "Quartz, Tibet", https://creativecommons.org/licenses/by-sa/2.5/legalcode

[ii] Page 2, Figure 1: Rock crystal rhyton archmus Heraklion, 1500 -1450 BC, Photo by anonymous. (https://commons.wikimedia.org/wiki/File:Rock_crystal_rhyton_archmus_Heraklion.jpg), https://creativecommons.org/publicdomain/zero/1.0/legalcode

[iii] Page 18, Figure 3: Doubly Terminated Quartz Crystal, Photo by Rob Lavinsky, iRocks.com – CC-BY-SA-3.0 (https://commons.wikimedia.org/wiki/File:Quartz-qtzrl-16b.jpg), „Quartz-qtzrl-16b", https://creativecommons.org/licenses/by-sa/3.0/legalcode

[iv] Page 21, Figure 5: Amethyst. Photo by JJ Harrison (https://jjharrison.com.au/) (https://commons.wikimedia.org/wiki/File:Amethyst._Magaliesburg,_South_Africa.jpg), „Amethyst. Magaliesburg, South Africa", https://creativecommons.org/licenses/by-sa/3.0/legalcode

[v] Page 27, Figure 7: Phantom Quartz, Photo by Rob Lavinsky, iRocks.com – CC-BY-SA-3.0 (https://commons.wikimedia.org/wiki/File:Quartz-139132.jpg), „Quartz-139132", https://creativecommons.org/licenses/by-sa/3.0/legalcode
Locality: Blue Phantom Mine, Mt Ida, Montgomery County, Arkansas, USA

[vi] Page 31, Figure 8: Quartz var. Citrine, Parent Géry (https://commons.wikimedia.org/wiki/File:Citrine_1_(Russie).jpg), „Citrine 1 (Russie)", marked as public domain, more details on Wikimedia Commons: https://commons.wikimedia.org/wiki/Template:PD-self

[vii] Page 38, Figure 10: UCL Museums & Collections (https://commons.wikimedia.org/wiki/File:Agate_in_UCL_Geology_Collections_1.tif), „Agate in UCL Geology Collections 1", https://creativecommons.org/licenses/by/3.0/legalcode

[viii] Page 48, Figure 11: Opal from Yowah, Queensland, Australia. Photo by Noodle Snacks (https://commons.wikimedia.org/wiki/File:Opal_from_Yowah,_Queensland,_Australia_2-2.jpg), "Opal from Yowah, Queensland, Australia 2-2", https://creativecommons.org/licenses/by-sa/2.5/legalcode

[ix] Page 57, Figure 14: Moonstone, Oslo, Norway. Photo by Lech Darski (https://commons.wikimedia.org/wiki/File:Kamień_księżycowy_(moonstone)_-_Larvik_k._Oslo._Norwegia.jpg), https://creativecommons.org/licenses/by-sa/4.0/legalcode

[x] Page 64, Figure 15: Microcline var. Amazonite, Photo by Parent Géry (https://commons.wikimedia.org/wiki/File:Amazonite,_quartz_300-3-7927.JPG), „Amazonite, quartz 300-3-7927“, https://creativecommons.org/licenses/by-sa/3.0/legalcode

[xi] Page 72, Figure 18: New Caledonia; Battle axe; Nephrite Jade
Photo by anonymous (https://commons.wikimedia.org/wiki/File:MET_02_18_297_S1_sf.jpg), „MET 02 18 297 S1 sf“, https://creativecommons.org/publicdomain/zero/1.0/legalcode

[xii] Page 75, Figure 19: Epidote in matrix with manganaxinite, Photo by Ivar Leidus (https://commons.wikimedia.org/wiki/File:Epidote_-_Canta,_Lima,_Peru.jpg), https://creativecommons.org/licenses/by-sa/4.0/legalcode

[xiii] Page 85 Figure 20: Azurite and Malachite. Photo by Marie-Lan Taÿ Pamart. Given by J. Pierpont Morgan in 1903 to the Gallery of Mineralogy and Geology of the Museum of Natural History, Paris. (https://commons.wikimedia.org/wiki/File:Malachite_and_azurite_Morenci_MNHN_Minéralogie_n1.jpg), https://creativecommons.org/licenses/by/4.0/legalcode

[xiv] Page 103, Figure 21: Topaz Cluster, Photo by Rob Lavinsky, iRocks.com – CC-BY-SA-3.0 (https://commons.wikimedia.org/wiki/File:Topaz-27da.jpg), „Topaz-27da“, https://creativecommons.org/licenses/by-sa/3.0/legalcode

[xv] Page 104, Figure 22: Colorless topaz. Photo by Géry Parent (https://commons.wikimedia.org/wiki/File:Topaz_24.jpg), „Topaz 24“, marked as public domain, more details on Wikimedia Commons: https://commons.wikimedia.org/wiki/Template:PD-self

[xvi] Page 105, Figure 23: Blue Topaz. Photo by Rob Lavinsky, iRocks.com – CC-BY-SA-3.0 (https://commons.wikimedia.org/wiki/File:Topaz-153562.jpg), „Topaz-153562“, https://creativecommons.org/licenses/by-sa/3.0/legalcode

[xvii] Page 108, Figure 24: Chrysoberyl in a star form. Photo by Matteo Chinellato (https://commons.wikimedia.org/wiki/File:Chrysoberyl-282796.jpg), „Chrysoberyl-282796“, https://creativecommons.org/licenses/by-sa/3.0/legalcode

[xviii] October 1906

[xix] Page 116, Figure 25: Elbaite Tourmaline, Photo by Rob Lavinsky, iRocks.com – CC-BY-SA-3.0

(https://commons.wikimedia.org/wiki/File:Tourmaline-51112.jpg), „Tourmaline-51112“, https://creativecommons.org/licenses/by-sa/3.0/legalcode

[xx] Page 118, Figure 26, Polychrome Tourmaline. Photo by Sailko, (https://commons.wikimedia.org/wiki/File:Tormalina_policroma,_da_madagascar_o_namibia,_01.JPG), „Tormalina policroma, da madagascar o namibia, 01“, https://creativecommons.org/licenses/by/3.0/legalcode

[xxi] Page 158, Figure 27: Baltic Sea Amper, Photo by M. Porsina, (https://commons.wikimedia.org/wiki/File:Балтийский_янтарь._Старение_янтаря..jpg), https://creativecommons.org/licenses/by-sa/4.0/legalcode

[xxii] Page 162, Figure 28: Soft Red Coral with Polyps, photo by Nhobgood Nick Hobgood (https://commons.wikimedia.org/wiki/File:Scleronephthya_sp._(Red_soft_coral_with_red_polyps).jpg), „Scleronephthya sp. (Red soft coral with red polyps)“, https://creativecommons.org/licenses/by-sa/3.0/legalcode

[xxiii] Page 164, Figure 29: Mother of pearl with accrued pearls. Photo by Hannes Grobe/AWI (https://commons.wikimedia.org/wiki/File:Mother-of-pearl_internal-beads2_hg.jpg), „Mother-of-pearl internal-beads2 hg“, https://creativecommons.org/licenses/by/3.0/legalcode

Made in the USA
Monee, IL
30 April 2026

49222808R00114